THE CITY

ENGLISH URBAN LIFE 1776–1851

HISTORY OF THE CITY

ENGLISH URBAN LIFE 1776–1851

JAMES WALVIN

Routledge
Taylor & Francis Group
LONDON AND NEW YORK

First published in 1984

This edition published in 2007
Routledge
2 Park Square, Milton Park, Abingdon, Oxon, OX14 4RN
Simultaneously published in the USA and Canada by Routledge
711 Third Avenue, New York, NY 10017
Routledge is an imprint of Taylor & Francis Group, an informa business
Transferred to Digital Printing 2007
First issued in paperback 2013

British Library Cataloguing in Publication Data
A CIP catalogue record for this book
is available from the British Library

ISBN13: 978-0-415-41757-0 (volume hbk)
ISBN13: 978-0-415-41933-8 (subset)
ISBN13: 978-0-415-41318-3 (set)
ISBN13: 978-0-415-86038-3 (volume pbk)

Routledge Library Editions: The City

English Urban Life 1776–1851

James Walvin

Reader in History, University of York

Hutchinson

London Melbourne Sydney Auckland Johannesburg

Hutchinson & Co. (Publishers) Ltd

An imprint of the Hutchinson Publishing Group

17–21 Conway Street, London W1P 6JD
and 51 Washington Street, Dover, New Hampshire, 03820, USA

Hutchinson Publishing Group (Australia) Pty Ltd
16–22 Church Street, Hawthorn, Melbourne,
Victoria 3122

Hutchinson Group (NZ) Ltd
32–34 View Road, PO Box 40–086, Glenfield, Auckland 10

Hutchinson Group (SA) (Pty) Ltd
PO Box 337, Bergvlei 2012, South Africa

First published 1984

Set in 11 on 13 point Garamond by Spire Print Services Ltd,
Salisbury, Wilts

Printed and bound in Great Britain by
Anchor Brendon Ltd,
Tiptree, Essex

British Library Cataloguing in Publication Data

Walvin, James
English urban life 1776–1851.—
(Hutchinson social history of England; 4)
1. Cities and towns—England—History
2. England—Social conditions—18th
century 3. England—Social conditions
—19th century
I. Title
942'.009'732 HT133

Library of Congress Cataloging in Publication Data

Walvin, James.
English urban life, 1776–1851.

(Hutchinson social history of England)
Bibliography: p.
Includes index.
1. England—Social conditions—18th century.
2. England—Social conditions—19th century. 3. Cities and towns—England—Growth. I. Title. II. Series.
HN398.E5W35 1984 306'.0942 84–12941

ISBN 0 09 156150 7 cased
0 09 156151 5 paper

Contents

Preface

It has become customary for authors to thank those people directly helpful in the writing of a book. In this case such thanks are difficult because this book has been shaped in large part by that extraordinary outpouring from social historians over the past twenty years. I am happy however to single out Edward Royle not merely for allowing me access to his books but for providing an invaluable critique of an early draft of this book. My thanks are also due to Ann Wong and Hazell Haines who typed the final manuscript.

In becoming a professional historian I was guided by a series of happy accidents. But if any one set of circumstances conspired to encourage me to pursue a historical career it was above all my school days in Manchester in the 1950s. At Ducie Technical School I was extraordinarily fortunate to be taught by a number of men whose encouragement and help edged me towards history. John Lee taught me most of the history I knew up to the age of 18. In Sam Hughes I was lucky to have a Headmaster who was also a historian. He shaped my progress in a number of ways, lending books, conjuring forth free dinner tickets at crucial moments and generally encouraging me to study history. It is only after teaching for a number of years that I have come fully to appreciate the importance of the efforts of those two devoted schoolmasters. To them and to my other teachers in those years I dedicate this book and hope that they see in it some of their own work.

J.W.
York
November 1983

Introduction

The modern observer would have had great difficulty in recognizing many of the physical features of mid eighteenth-century England. This would have been much less true a century later. In the intervening years a series of profound changes had transformed not merely the nature of English economic and social life, but had even altered the physical face of England itself. Indeed it is thought that these physical changes were unequalled by any others in the entire millennium since the Anglo-Saxon settlements.[1*] It is true that many of these originated long before these years. None the less it was in the late eighteenth and early nineteenth centuries that they reached new levels. Land enclosure, land reclamation, the draining of marshes and fens, the gradual erosion of woodlands, improvements in and proliferation of roads and the inauguration of a major network of canals – later the railways – all these are well known and familiar and cumulatively helped to reshape the face of England. But the most dramatic, far-reaching and ubiquitous of all these changes was the increase in the number and size of towns and cities. Again, the development of the English town had been a feature of early eighteenth-century history, but the pace and the scale of urban growth quickened perceptibly in these years (and even more so in later years). Moreover, so rapid did the growth of towns become that they rapidly outstripped the ability of contemporaries adequately to regulate and control that growth, and all its attendant problems. Whereas there had been notable efforts in the early eighteenth century at early town planning (notably in the spas), the years under review here marked a contrary trend; of the apparent helplessness of man to control the growth or to direct the nature of contemporary urban growth. One of Goldsmith's characters commented in 1773, 'In my Time

[*]Superior figures refer to the References beginning on p. 197.

the follies of the Town crept slowly among us, but now they travel faster than a Stage-Coach.'[2]

What follows is not an attempt to describe the process of urbanization in these years but to assess the impact of town life upon the swelling armies of town dwellers. Contemporaries had long been conscious that the sensibilities of town dwellers were quite different from those of rural people. And as an increasing proportion of the population found themselves living in an urban habitat, it seemed clear to many that the dominant sensibilities of the English were urban. In large part this was regarded as a vice rather than a virtue, as a visiting Swiss industrialist confided in his diary in 1851,'Big cities, with few exceptions, ruin the people who have to spend their lives in them.'[3] None the less, whether for good or ill, there developed among the English people an unmistakable 'urban identity',[4] an identity forged partly by the unavoidable and capricious accidents of town life and partly by the conscious responses of individuals and groups to the difficulties of the urban environment. In many respects the English town dweller could do little about the material conditions (and problems) of urban life; of housing, work and wages, food and physical well-being. Yet the very gathering together of people in close urban circumstances also made it possible – as rarely before – to meet, organize and to lobby to promote political or economic interests.

A word needs to be said about chronology. There is an inevitable artificiality about the way historians slice up the processes of history into neat chronological chunks. The need to impose retrospective sense and order on the confusion of events in past time is, however, basic to the historian's task. I am conscious of the risks involved in such exercises of 'periodization', and in choosing the years 1776–1851 I do not wish to convey the impression that they embrace a perfectly coherent and autonomous period. This study spans the years between the outbreak of the American Revolution – the origins of the modern democratic tradition – and the point at which the mid nineteenth-century English were shown to be an urban people. The terminal date also marks the onset of newly-won social tranquility, and the material achievements of the embryonic industrial nation – celebrated at the 1851 Great Exhibition. It is a relatively brief period – no more than a mature lifetime – but in those seventy-seven years the nation underwent a process of change quite unlike anything before. Moreover, for twenty-seven of those years the nation was at war, though it could be argued that war was almost the 'typical' eighteenth-century English condition. For our purposes, however, we need to remind ourselves that many of the difficulties created by the economic and demographic changes of contemporary life were compounded by the human and financial ramifications of warfare.

The book which follows is an attempt to sketch in broad outline some of the major objective changes shaped by the aggregation of ever more people into an urban environment, and the efforts of those people to accommodate themselves, in most aspects of their lives, to the urban world. As will become clear it is quite impossible to draw a sharp line between rural and urban life; the one was mutually linked to the other. None the less the focus of this book is on town life; rural life is a parallel phenomenon (dealt with in a companion volume) which is discussed only at those points which illustrate or determine town life itself. There are, however, historical distinctions which divide the rural from the urban world; the book which follows addresses itself to a number of those distinctions.

Perhaps the most ambitious aim is to seek to illustrate not merely the obvious changes across time but also to discuss the continuities. It has become something of a historical caricature to imagine that the process of urban growth (and industrialization) swept all before it; that the English urban people were effectively deracinated from their rural and pre-urban past. In some respects, the demographic changes (and even the conscious efforts of certain English people) were undoubtedly conducive to that end. But there was a host of ways in which English society remained 'traditional'; in which English people, living in towns and cities, conducted their daily lives in ways which would have been instantly recognizable to their forebears. Of course, to describe such continuities alongside the parallel changes involves striking a balance. But if I have opted for any particular stance it is to stress the continuities in English social life. I have done this partly because this has been the thrust of my own work in a number of earlier books. But I have also consciously set out to explore what I regard as some of the key continuities in English social life (in popular politics, recreations, rituals and family life for instance), in order to provide an antidote to the contrary impression cultivated by numerous other studies which portray the English people turning their collective backs on their pre-modern past. Thus, the book addresses itself not to urban history *tout court* but to many of the formative experiences of working people – the common people – coming to terms with the objective (and generally harsh) realities of urban life; often helpless in the face of uncontrollable elements, sometimes resistant to difficulties, dogged in the defence of traditional values and behaviour, though often times succumbing to material dangers which were themselves new and awesome. In many respects these were transitional years; a phase in the protracted transformation to an even more acutely urbanized and industrialized society. In that sense it is a story that will be continued by subsequent volumes in this series. Let us begin however by considering the changes in the lives of English people brought about by the remarkable growth of towns.

Part One

A Changing Nation

The most striking changes in English life in these years were the obvious physical changes, notably the growth of towns. But beneath the rather bland generalizations commonly proffered about 'urban growth' lay a rich mosaic of local distinctions (which survive of course to this day). And yet even allowing for local peculiarities, there were distinct problems facing town dwellers throughout the nation – especially low income groups – few more troublesome than the practical difficulties of town life – of housing and home, of food, nourishment and of clothing. Of course all of these were determined in large part by the nature of work and by the ability (or inability) of the individual and family to earn a consistent and adequate wage to pay for life's essentials. Even when in work – and adequately rewarded – the town dweller was exposed to a range of physical dangers, none more acute, persistent and damaging than contemporary illness and physical suffering. Furthermore, as the nation became more urbanized, many of these medical problems became more acute and widespread. Indeed so many seemed to be a function of, or at least to thrive in, the urban habitat.

In retrospect one of the key changes in English life in these years was not physical or material. The distinctions of social class, so inescapable a feature of subsequent social and political life in England, developed in these years. Precisely how this happened continues to cause historical controversy. In this book the process is given a central role, alongside (and in part a function of) the physical transformation of the nation.

1 *People and towns*

English people throughout the late eighteenth century were regularly impressed by the population and urban growth taking place around them. Looking back a mere decade from 1791 William Hutton wrote of Birmingham that

> Streets and houses have arisen, on our right, out of solitary fields. The cattle have been turned out of their pasture to make room for man, and the arts are planted where the daisy grew. These additions are so amazing, that even an author of veracity will barely meet belief. A *city* has been grafted upon a town.[1]

The sense that urban growth was spitting out bricks and buildings and roads across the land was commonly repeated, most graphically perhaps by a generation of cartoonists and caricaturists who expressed, often in grotesque form, the transformation at work around them. George Cruickshank's cartoon, 'London going out of town' (1829) portrayed the ruthless and remorseless advance of jerry-building; rural Hampstead under bombardment from a stream of bricks shot forth from the brick kilns. And the whole startling scene was shrouded in palls of black industrial and domestic smoke. It was a hellish image of a landscape transformed and destroyed by the advance of a human tide.[2]

The same impression was to be found wherever one looked, though with local peculiarities and a different pace of change from town to town. The physical appearance of the urban environment was embroiled in a process of unprecedented change to cope not merely with the transformations in economic activities (themselves varying from one town to the next) but, most crucially perhaps, with the massive upsurge in population. In retrospect it is possible to plot these changes in a number of ways but whichever tack we choose, the impression is the same, and simply goes to confirm the astonishment felt by contemporaries at what was happening to England.

The most seminal – and obvious – of all transformations was the growth in population. In the years covered by this book the population of England and Wales expanded from a little over 7 million people to something under 18 millions. Naturally enough growth of this order traces its origins to an earlier period. From the early eighteenth century there had been a gradual but perceptible increase in population; but we need to remind ourselves that the English population of 1700 was a mere 5 million souls, 80 per cent of whom lived in the countryside. From the mid century, however, population increase began to accelerate, reaching unknown levels in the 1780s and 1790s, continuing into the nineteenth century and 'peaking' in the middle of the century.[3] Of course it is also true that, with the painful exception of Ireland, a similar pattern repeated itself (with regional variations) throughout most of western Europe. Indeed it is only towards the end of the eighteenth century that population growth in England and Wales began to deviate from the rest of Europe, to outstrip our European neighbours and become a by-word for a society beset by population expansion – with all its attendant problems.

The key historical question remains; precisely why did population growth become so marked? Modern researchers seem generally agreed that the central and determining influences were the changes in the birth rate. Recent calculations show that it was the upswing in birthrates which accounted for perhaps 70 per cent of the increase in population in these years.[4] There were, however, remarkable variations within this generalized experience of population growth. Most marked of all was that between agricultural and industrial areas. Wherever buoyant industrial activity was to be found — even in rural setting — local population growth greatly outstripped the population increase to be found in agricultural settings. The end result – and one which became ever more pronounced in the late nineteenth century – was the geographical shift in the English population. As industrial activity itself became an increasingly urban phenomenon, population growth in those towns and cities rapidly outstripped its rural and agricultural hinterlands.

In these years the preponderance of the population became increasingly urban; gradually but perceptibly the English were transformed into a nation of town dwellers. In the late seventeenth century, the proportion of the population living in urban areas was perhaps 1 in 4. By the middle of the nineteenth century it was 1 in 2. It has been calculated that in 1801, 33.8 per cent of the population of England and Wales could be classified as urban. With each successive decade this percentage increased.

1811 – 36.6 per cent
1821 – 40.00 per cent

1831 – 44.3 per cent
1841 – 48.3 per cent
1851 – 54.0 per cent[5]

Yet even at the end of our period it is tempting to overstate the degree of urban domination of English life. Even in the greatest of English cities – London, whose population in 1851 was 2.3 million – it remained easy and quick to walk to the surrounding countryside. When, in the course of the 1790s, the radical working-men's organization, the London Corresponding Society, called three large-scale public meetings they were convened in accessible open fields to the north of the city, in Chalk Farm, Islington and St Pancras. Fifty years later, when new cemeteries were developed on the edges of London, they were opened in areas which, today, seem remarkably central; in Highgate and Kensal Green for instance. Furthermore, and as we shall see later, the rhythms and pulse of agricultural life continued to play a powerful role in the economy and social life of London right up to (and beyond) the mid nineteenth century. Many of the fairs and festivals of London were not only traditional and often very ancient but were primarily rural occasions, held either on the edge of the capital or imported into its famous and communal public places. A similar tale could be told of those northern towns and cities which came to characterize the problems (and potential) of urban and industrial life. In Oldham for instance despite its massive population growth, its textile proletariat continued to enjoy the boisterous customs and pleasures of their pre-industrial forebears until, in 1849, the new local police force imposed a new and unwelcome moderation on the proceedings. When in 1848 Elizabeth Gaskell wrote *Mary Barton*, she began with these words:

> There are some fields near Manchester, well known to the inhabitants as 'Green Heys Fields', through which runs a public footpath to a village about two miles distant.

Thus, only half an hour's walk from Manchester, there were to be found

> an old black and white farm house, with its rambling outbuildings. . . . Here in their seasons may be seen the country business of hay-making, ploughing, etc., which are pleasant mysteries for the people to watch; and here the artisan deafened with the voice of tongues and engines, may come to listen awhile to the delicious sounds of rural life: the lowing of cattle, the milk-maids' call, the clatter and cackle of poultry in the old farmyards. You cannot wonder, then, that these fields are popular places of resort at every holiday time. . . .[6]

It is then in many respects impossible to neatly separate the two; to suggest that there was a sharp dividing line between the rural and urban worlds. It was more like a spectrum of social experiences, clearly very distinct at the far edges but becoming blurred and then indistinguishable at the points of contact. After all, a very substantial proportion of all town dwellers had been born in the countryside. Notwithstanding these qualifications, however, England became progressively more urban throughout our period if only because an ever-increasing proportion of the people moved to or were born in towns and cities. Put at its simplest, it was in these years that England began to transmute itself into an urban society; a society marked by the quantitative shift in its population. But, equally crucial, this involved a qualitative change, for English people were becoming a nation whose seminal and most significant experiences were being forged in an urban habitat – even allowing for the common, ancestral – and continuing – cultural ties with rural life.

England was transformed in these years not merely by the growth of population but also by a marked and continuing shift in its location. This was at its most obvious in the proliferation and growth of towns and cities. In the early eighteenth century, the pattern of towns was basically that inherited from the late Middle Ages. There were between 500 and 600 small market towns with populations of between 500 and 1800 people. Among their number there were certain prominent towns – often 'country towns' – which provided a legal, administrative or ecclesiastical centre for its immediate rural region. But above all others there stood the five 'provincial capitals' of Norwich, Bristol, York, Exeter and Newcastle, each of which lay at the heart of a web of local economic activity, communications and social life. Equally important, all of these cities were distant from London (whose population was about twenty times that of its next rival). This pattern of urban hierarchy which had remained stable for centuries was to be utterly overturned within a relatively short period. By 1801 only Bristol could be counted among the nation's top five biggest provincial towns. Norwich was seventh, Newcastle thirteenth, Exeter fourteenth, and York sixteenth (Table 1). All had been swiftly relegated by the burgeoning industrial towns and ports. Yet it needs to be stressed that these 'new towns' had themselves blossomed from well-established, and in some cases ancient, urban communities.

As towns grew in number and size, their importance to the rural world was also enhanced. Increasingly, they came to provide rural life with the highly specialized skills which, though urban in origin and nature, were crucial for the parallel changes taking place in rural society. It was from the towns that the new scientific, financial and technical ideas seeped out into the countryside. It seems clear for instance that much of the improvement in agricultural life in our period could be traced to the

Table 1 *Largest towns in England and Wales (London excluded)*

	1801		*1851*
Manchester	84,000	(including Salford)	367,955
Liverpool	78,000		375,955
Birmingham	74,000		232,841
Bristol	64,000		137,328
Leeds	53,000		172,270
Plymouth	43,000		
Norwich	37,000		
Bath	32,000		
Portsmouth	32,000		
Sheffield	31,000		135,310
Hull	30,000		
Nottingham	29,000		
Newcastle	28,000		
Bradford	–		103,310[7]

educational, cultural and intellectual transformations which we associate with towns.[8] The Literary and Philosophical Societies which became such an intellectual feature of the newer industrial centres from the late eighteenth century onwards often addressed themselves to practical and scientific issues of great relevance and importance to rural as much as urban life.

Perhaps the most popular of images of the early Victorian town and city – conveyed by graphic and literary commentaries from the 1830s onwards – is that of a filthy, overcast and unhealthy industrial centre. Indeed, it is a common but misleading assumption to think that urban areas were of necessity industrial centres. In the early eighteenth century, however, many major English industries thrived in a *rural* environment. Even by the late eighteenth century maps clearly show the very small degree of urban development in areas where industry was expanding. And yet within two generations this picture was to be utterly transformed as the people of England and the industries in which they laboured became increasingly concentrated in urban areas. When towns began to grow in our period, the single most important influence over that growth was the development of industry; conversely the decline of older industries (textiles in East Anglia for instance) retarded the physical and human expansion of local towns. Towns withered or expanded in relation to the health and buoyancy of their basic trades and industries. The Lancashire textile towns provide perhaps the clearest example of this point, growing dramatically in direct proportion to the expansion of local textile industries. Whereas the population of Lancashire in 1751 was 318,000, by 1801 it was 694,202 and 1 million by 1821.[9] In the process Manchester became the natural centre of local textile transactions, its own population of course growing dramatically to more than one-third of a million by mid century. Other Lancashire towns gradually succumbed to the gravitational pull of Manchester's banking, warehousing, distributive and negotiating strength. Within Manchester there developed a remarkable range of specialized skills and trades, to satisfy the economic needs of the city itself and of its expanding industrial heartland – a classic illustration of the economic specialization associated with urbanization.[10]

A similar process was at work in the Yorkshire woollen industry where Leeds emerged as the pre-eminent local city through which (like Manchester) raw materials and finished goods were channelled. And so too were the agricultural products from the wider rural interior. Both cities became the economic and social hub for the complex local wheel of industry.[11] This pattern – of urban development attendant upon major industrial change – could be found throughout late eighteenth- and early nineteenth-century England. It happened in Nottingham and Leicester (in the hosiery industry) and also in the metal industries of the Black Country, where the older artisanal skills in metal were revitalized and transformed by improvements in communications and the accelerating demand for local products and skills.

British maritime ascendancy in the eighteenth century and the periods of international warfare, first against the Americans and later the French, provided a major stimulus to the urban growth of the nation's ports and dockyards. Liverpool, soon to become the major entrepôt of the textile trades, had emerged as an important port in the second half of the eighteenth century thanks to the rapid growth of its trading links to Africa, North America and the Caribbean. By 1770 Arthur Young thought Liverpool 'too famous in the trading world to allow me to pass it by'.[12] Although it is true that a great deal of English domestic trade was seaborne, it was the massive growth of the British empire and of international trade in the eighteenth century which so helped to stimulate the development of the ports. This was clearly the case with Bristol and Liverpool but also (and pre-eminently) with London. The ports of Hull and Newcastle similarly benefited from the expansion of overseas trade. But these and other ports also needed an industrial hinterland to feed their international or domestic trading links. Ports with vital and expansive hinterlands and lucrative overseas connections grew rapidly as urban areas. But those whose interior economic and trading activities declined found their own physical growth arrested or in relative decline.[13]

To expand as urban areas, ports needed a buoyant hinterland economy; Whitehaven, for instance, lacked this essential and hence went into decline from the late eighteenth century onwards. Some ports were unable to grow because their basic source of maritime trade was too limited, not sufficiently varied, to generate inland economic activity. It is, for instance, claimed that it was Liverpool's broad trading base which enabled it to expand so dramatically in the late eighteenth century. Furthermore, Liverpool's eventual eclipse of Bristol was aided in large part by the opening of canals which fed goods from the Midlands more quickly and cheaply to Liverpool than could be done to Bristol.

The thriving ports were, then, bustling urban areas. Some were elegantly planned, characterized by up-to-date port facilities, fashionable homes and business premises for the families and companies which financed and profited from the ports' good fortunes. Many developed a distinct and often fashionable social life which, if not on a par with the *élan* of the spa towns, was none the less appealing. They also inevitably housed their own social problems; prostitution and poverty, overcrowding and dirt – all (and more) seemed unavoidably attendant on thriving urban growth and economic development. It was from these hubs of international commerce that so many families of initially moderate means were able to elevate themselves from mere traders into dynasties of opulence and rural property; making that important transition from city traders to country gentlemen. It was of course a risky venture and many were reduced in circumstances. But the most abiding change brought about by their efforts was not so much in their own lives as in the emergence of major urban port facilities which came to dot the coastline of late eighteenth-century England. These were the very facilities which were to be so instrumental in tapping and then advancing the economic changes we think of as the industrial revolution.

The growth of industrial or maritime towns is not perhaps surprising. Yet it may seem odd that the most dramatic form of urban growth from the late eighteenth century onwards was to be found in the resort towns. To cater for the middle- and upper-class taste of 'taking the waters', at the spa, or, later, at the seaside, there emerged those resort towns able to provide an unusual range of specialized services and trades for the pleasure of their growing number of visitors. Bath remained the pre-eminent inland spa, a major town of 34,160 people in 1801. But it was the switch from spa- to sea-water in medicinal recommendations which was initially responsible for the rapid proliferation of seaside resorts. In 1801 Brighton's population was 7339, Scarborough's 6409. By 1821 Brighton's population had grown to 24,429 – an indication of the important but generally neglected fact that in the course of the first half of the nineteenth century it was the new resort towns which were to register the sharpest increase in population growth and urban development.[14]

Towns had traditionally provided a base for organized and commercial pleasures. Throughout the eighteenth century (and much earlier) London had been renowned – among other things – for its extraordinary array of pleasures available to high and low; its public parks, spas and pleasure gardens, its fairs and tea-gardens, its street entertainments, and its complex and ancient calendar of folk customs and ceremonies that punctuated and highlighted the working year. Much the same was true, though on a less spectacular scale, of the other 'provincial capitals' and many smaller towns. Long before our period, for instance, there had grown up that distinctive urban development of the spa towns where pleasure could be had in conjunction with the pursuit of good health. Epitomized by Bath, the spa towns were (and some remain) urban and architectural monuments to contemporary tastes and fads, enjoyed or endured by the wealthy. Built by a mixture of private and municipal capital, boosted by upper-class and aristocratic patronage, the spas, by and large, remained physically and financially remote from the needs and the interests of the great bulk of the people.

One consequence of the medical advice to turn to sea-water for the benefits allegedly bestowed by spa-water was to bring the habit of 'taking the waters' within the reach of the lower orders. It is characteristic that the particular towns to cater for more popular tastes – the seaside resorts – were to be in many respects the nineteenth-century equivalents of the older spas. But when we examine the broader history of both spas and seaside resorts we can detect another important feature of urban growth; the specialization of economic functions in particular towns and cities. And while the popular image of the mid nineteenth-century town is the specialized cotton, woollen, metal or mining town, we must not forget those towns that to this day continue in the economic tradition for which they evolved; the provision of collective pleasure on an unprecedented scale – beside the seaside. Blackpool, no less than Oldham or Sunderland, was a quintessentially nineteenth-century town.[15]

The English population in these years not only increased, it also moved. There was, throughout, a marked movement of people from rural to urban life – as of course there had been, on a much less significant scale, since time immemorial. London, for instance, could never have reached its massive pre-industrial size without the seductions it offered to distressed rural people. Throughout the sixteenth and seventeenth centuries urban growth, especially in London, had derived in large part from the wretched migrations of British people.[16] The pattern became ever more pronounced in these years. Population growth was a feature of rural no less than urban life, though at a lower rate of increase than in urban or industrial areas. Between 1751 and 1831 the population of purely rural counties, for instance, increased by 88 per cent, while that of the industrial and commercial counties rose by 129 per cent.[17] Large numbers of young rural people, men and women, drifted to the towns, normally in the first place to the nearest town, providing what has been described as 'the stock from which, in the eighteenth century, the large towns, London in particular, were able to sustain relatively rapid growth despite high mortality'.[18] Huge numbers of people moved permanently into the towns, though there were also large numbers who moved temporarily – or were indeed peripatetic between their rural homes and what was considered to be the greater economic opportunities of town life. Seasonal movements of labour, annual hirings of farm servants and migrant labour all drew the population away from the countryside. Among other things, this had the effect of forging close personal and family ties between urban and rural society; closer, indeed, than is often recognized. It also had the effect of making rural society itself more 'fluid' and less secure than is sometimes realized. The evidence however (as we might expect) is confusing. Those fast-growing industrial areas – Lancashire and Warwickshire for instance – which needed immigrants for their expansive labour force, tended to draw their migrant labour from short distances. Indeed, it has been calculated that migrants seeking work rarely travelled more than 10 miles from their place of origin.[19] Yet there is also a contrary impression; of men and women traversing long and arduous distances to seek work and security. Diaries and memoirs regularly record incidents of poor people begging at the door of the propertied, seeking refreshment, rest and warmth as they wandered across the land. The Wordsworths in their lakeland retreat recorded a number of such incidents. In March 1802, Dorothy Wordsworth noted 'a sailor, who was travelling from Liverpool to Whitehaven called, he was faint and pale when he knocked at the door. . . .'. Later that same week, 'A sailor begged here today, going to Glasgow.' Two months later, the Wordsworths encountered 'a woman with three children travelling from Workington to Manchester'.[20] Much later, at mid century in London, Henry Mayhew made the following significant comment,

> The nomadic races of England are of many distinct lands – from the habitual vagrant – half-beggar, half-thief – sleeping in barns, tents, and casual wards – to the mechanic on tramp.... Between these two extremes there are several

mediate varieties – consisting of pedlars, showmen, harvest-men, and all that large class who live by either selling, showing or doing something through the country. . . . Besides these, there are the urban and suburban wanderers, or those who follow some itinerant occupation in and round about the large towns.[21]

As we might expect, nowhere was the evidence of migrating and travelling people more striking than in London.

In the first half of the eighteenth century, more than half a million people migrated to the capital – a movement largely offset by high levels of mortality in London. And those who settled securely in the capital were in addition to those legions of rootless, travelling workers whose seasonal, migratory, casual work – or the search for it – brought them temporarily into the metropolis. These 'comers and goers', the 'migrating class', were to characterize London life throughout these years and beyond.[22] London had traditionally exercised a gravitational pull over the desperate or the ambitious throughout the British Isles and had for centuries provided a home for substantial numbers of non-local people. The strong and the willing, the enterprising along with the wretched and the maimed found their various ways to London to seek a livelihood. The capital seemed to hold out the prospects of work for the able bodied, and opportunities for begging for the miserable.

If any person is born with any defect or deformity or mained by fire or other casualty, or any inveterate distemper which renders them miserable objects, their way is open to London, where they have free liberty of showing their nauseous sights to terrify people, and force them to give money to get rid of them.[23]

London was home for substantial minorities from throughout the British Isles, notably the Welsh, Scots and Irish (all with their own social organizations). Even that most famous of eighteenth-century Londoners, Dr Johnson, whose aphorism about London is repeated to this day, was himself a provincial and was mocked for his Litchfield accent.

London was a cosmopolitan city, providing a temporary home for sizeable communities of Europeans and a permanent abode for others. The Irish were of course a famous London community. So too were the French Huguenots, European Jews and, by the late eighteenth century, those pockets of black slaves adrift in London from the expansive slaving empires that spanned the Atlantic. By the end of the American War (1783) the 'black poor' were so pressing a 'problem' that the government devised an expensive (and ultimately disastrous) scheme to 'repatriate' them to Sierra Leone.

Despite the obvious and ubiquitous poverty in London, people continued to settle in the capital, and the pace increased in the nineteenth century. In one decade alone, between 1841–51, one-third of a million immigrants settled in London, including 8000 from Scotland, 46,000 from Ireland and 26,000 from overseas. By 1841 the local Irish (82,291 of them) formed 3 per cent of the capital's entire population. By 1851 there were 109,000 Irish-born living in London.[24] And the Irish were of course famous for their poverty.

The poor in London were to be counted in their legions. Country girls in search of domestic work (often, however, forced to join the ranks of the capital's prostitutes); young men seeking apprenticeships (but finding instead abject destitution); these and many more formed the human aggregate of the nation's largest city – whose total population was perhaps three-quarters of a million by 1780. For many of them, crime became the only way of scratching a precarious living from the unyielding metropolitan environment. Unable to live within the law, large numbers were obliged to live outside it. Poverty was then an inescapable feature of metropolitan life – and had that peculiarly concentrated and therefore unpleasant feature which made it so different (in scale and quality) from poverty in the rural world. The poor, obviously, were to be found in abundance throughout the rural world, but poverty in its most taxing form was to be found in London (and other cities). With the further development of towns, urban poverty was to become a characteristic blight of English life on a new, nationwide and unavoidable scale.

There was another and no less significant aspect to the extraordinary movements of British people in these years. Long before this period, the British, like the Spaniards and Portuguese even earlier, had established a reputation for themselves as international emigrants on a major scale. Indeed new imperial and overseas settlements had, from the late sixteenth century onwards, been made possible by the flow of people away from the British Isles – to North America, the Caribbean and the Indian sub-continent. This was a pattern which was to continue and to increase through the years 1776–1851; it has been calculated that in this time some 845,000 people migrated overseas (Table 2). Clearly, this outflow of people affected the nature of population growth in England itself, but it was only a relatively minor factor when set against the overriding and determining fact of the surplus of births over deaths.[25]

The migrations overseas were to shape future relations of the metropolitan power with those far-flung regions of the world, notably the white colonies that offered a home for the migrating British. To a marked degree Britain's future was to be inextricably involved with her colonies and dominions; forging that special economic and political relationship with distant 'kith and kin' which was to prove in many respects as

Table 2 *Migration overseas (per decade)*

1771–81	48,342
1781–91	32,642
1791–1801	40,855
1801–11	61,863
1811–21	91,117
1821–31	140,495
1831–41	191,785
1841–51	238,746

troublesome as it was beneficial. We can only speculate on what might have happened in Britain had these armies of people not been driven (or attracted) to seek a better life overseas. For those who stayed, however, there remained that complexity of difficulties and dangers which in many respects became more serious, more daunting and (for two generations at least) apparently more intractable as ever more English people struggled to secure a living in the uncertain and unquestionably squalid circumstances of urban life.

Most worrying of all features in the growing towns of England were the conditions under which their expansive labouring population were obliged to live. No one familiar with the home circumstances of rural labourers in the pre- or post-industrial world would argue that the homes of rural people were any other than squalid. There was, it is true, an abundance of literary and artistic romanticizing about the happy homes and cheerful hearths characteristic of labouring England. The reality, however, was quite different, and getting progressively worse as ever more people lived in towns. This was true of all towns – as it had been for centuries of London – but it was seen in its starkest and cruellest form in the buoyant industrial towns. Indeed, in many of the industrial towns it was the local industries that made domestic life so much more offensive; unregulated workplaces, cheek by jowl with housing, belched out their smoke, their waste and their noise.

Manchester was infamous by the early nineteenth century, but even as early as 1795 its problems for the local labouring force were unmistakable. John Aiken wrote in that year,

> As Manchester may bear comparison with the metropolis itself in the rapidity with which whole new streets have been raised, and its extension on every side towards the surrounding country; so it unfortunately vies with, or exceeds, the metropolis, in the closeness with which the poor are crowded in offensive, dark, damp, and incommodious habitations, a too fertile source of disease.

True or not, contemporaries firmly believed that the appalling living conditions, close to open sewers, primitive privies or dunghills, were the causes of illness throughout the towns.[26] Yet Manchester was only the most shocking example of a phenomenon that came to characterize plebeian domestic life throughout urban England.

In many respects, the poor housing of early urban (especially industrial) England was predictable. Wherever rural industries existed in the eighteenth century there were to be found abject domestic circumstances; dirty, damp, overcrowded and unhygienic homes. This was particularly true of mining villages. But even those abject conditions were seen to be overwhelmed – in scale, intensity and consequence – by the dramatic growth of the English towns with their core of terrible housing. The age-old urban problems of London were rapidly replicated across the country.

There were, it is true, some notable attempts at orderly town planning and the provision of decent plebeian housing. But planners and financiers were simply overwhelmed by the human tide seeking accommodation. With no real legislative or bureaucratic controls, town growth rapidly created a mosaic of habitations, different from town to town, whose pervading wretchedness was ubiquitous. Old towns found themselves squeezed with more and more people as densely-crowded accommodation intruded into available areas, streets and alleys. In Leeds and Nottingham the population, initially, was packed into the existing city areas, doubling (or worse) the human density. Elsewhere, of course, new (often 'jerry-built') housing spilled out into the encircling fields, normally close to places of work, generally densely crowded and, almost invariably, quite devoid of the basic facilities of water, sewage, lighting or ventilation. It is then not merely housing *per se* but that complex of social squalor and deprivation – at the heart of which lay the overcrowded home – which seemed to weave a web around urban growth throughout England in these years. There were great differences between the types of plebeian homes throughout urban England, but they shared enough distressing similarities to enable us to group them together.[27]

Above all else they stank. Time and again, outsiders – foreign visitors, doctors, social or political commentators – remarked on the stench of the English towns. And in a society which believed in the miasma theory of disease this was to have important consequences for arguments about medical and urban amelioration. In the words of one loquacious Irishman living in Leeds in 1840 'the smell was bad enough to raise the roof off his skull'.[28] They were, obviously, also overcrowded. No matter how big the physical expansion of the towns, the plebeian streets, alleys and courtyards were always teeming with people. And it took no medical training to suspect that such overcrowding (the figures different from one town to another) in such deprived circumstances might lead to disease and infirmities. But for the individual families obliged to endure such conditions, and their numbers increased rapidly (relatively and absolutely) in these years, the problems of housing reduced themselves to the simple determinants of their daily lives. In the words of one historian of working-class housing, the working-class family had 'not enough space, not enough warmth, not enough light, not enough furniture'.[29] Many years later, when domestic conditions for many working people *had* improved, Henry Price, a cabinet maker, poured scorn on the eulogies of the 'happy home'.

> The Merry Homes of England Around the fires by night. Someone has sung about them. But they could not have known much about them. The vast majority of them in the Towns and Cities have no room to be merry in.[30]

What needs to be stressed is that for literally millions of English people in these years, the formative domestic experiences – of love, marriage, birth, family, suffering and dying – were shaped by the unquestionably impoverished conditions of the urban home. As ever more people lived in towns, these wretched domestic conditions provided the broad outlines of personal, social and psychological development for an increasing proportion of the English people. When we look at working-class homes and working-class quarters of English towns in these years and when we consider the deterioration in their nature and condition, brought about among other things by the complex concentration of population, it is hard to see a glimmer of that material improvement claimed by some historians. Towns were naturally complex economic organizations, but they were, first and foremost, the home for ever more people; their basic unit was the family-based home. And for the non-propertied, the labouring people, the great majority able to do little about their own fate, these homes were, throughout these years, bleak and forbidding. Time and again, nineteenth-century working-class autobiographers remarked on this fact. From that source alone (and there is of course an abundance of parallel confirmation), one recent historian has concluded,

> The home was seen not as a refuge but as a cockpit, the arena in which the consequences of exploitation and inequality were experienced and battled with.[31]

Thus, when we break down the residential quarters of the town into their fundamental atomic part – the home – it is perhaps easier to appreciate why so many people (working class and their betters) were in no doubt that the march of the town, the transformation of the English people into a nation of town dwellers, was a marked turn for the worse. It was to be well beyond the limits of this study, before the material benefits of industrial expansion, in conjunction with effective cleansing and purging of the urban habitat, began to yield significant personal and family improvements for ordinary labouring people. If we need further evidence for the baneful consequences of town life upon English people in these years, we need only turn our attention to the question of their ailments and diseases.

2 *Suffering people: illness and death in the urban world*

It may seem obvious, or at least commonsensical, that the population growth which dominates English history in these years was in some way a reflection of improving material and social conditions. Closer investigation, however, suggests a less predictable image, one, furthermore, which has attracted unusually intense historical research for many years. The new demographic patterns seem clear enough, however. Stated crudely, the population began to rise, in the short term, because of a fall in the death rate that was paralleled by a high birth rate. The number of live births per thousand of the population greatly (and consistently) outstripped the numbers of deaths per thousand.[1] But at the heart of the problem of population growth lay the complex morphology of disease and the disease environment. Indeed, the changes in contemporary diseases and people's immunity to them is often thought responsible for the fall in the death rate from the late eighteenth century onwards. This appears to be especially true in the case of infectious diseases, notably TB, typhus and smallpox.[2] We need only recall the frequent and periodic devastation caused by plagues in earlier periods – the Black Death or the mid seventeenth-century plague, for instance – to appreciate their absence in more modern times.[3] Pre-modern Europe had been attacked, in an almost cyclical fashion, by catastrophic failures of harvest and/or ravages of major diseases, resulting in an arrest or even a decline in population growth.[4] Yet, it remains uncertain exactly why the particularly virulent diseases ceased to wreak their traditional havoc among western societies. It is possible, for instance, that improvements in climatic conditions, which made themselves felt in the late eighteenth century, helped to raise the levels of human resistance to certain diseases. Other more immediate and more tangible factors present themselves as more attractive explanations.

It has long been claimed that medical improvements made appreciable inroads into the traditional levels of mortality. Recent work, however, suggests that, to the contrary, medical changes, with a few notable exceptions, made little impact before the mid nineteenth century. It is now thought doubtful that changes – in midwifery, new drugs, vaccination, and even the establishment of new hospitals and dispensaries – made much appreciable impact on mortality. Similarly, the general improvement in personal hygiene – much emphasized in earlier histories – did not begin to make itself felt much before 1860, and could hardly have contributed to the earlier fall in mortality. Indeed, the appalling urban conditions which form a theme to this book and which, if anything, were worsened by the rapid spread of the urban environment, were the very factors blamed by contemporaries for the apparently non-irradicable human problems of urban England. There were, it is true, some tangible improvements in these years. More and more people used soap, and many more of their clothes were (thanks to the outpourings from the textile industries), easily washed and cheaply replaced. Yet, all this is likely to have been merely marginal to the main impact of disease and its effects on the overall mortality rates.[5]

Changes in the nation's diet and eating patterns might also be thought to provide scope for explanation of falling mortality rates. And yet, even here the historian faces confusing evidence. While it is undoubtedly true that agriculture became more efficient and widespread, there remained a persistent complaint throughout these years that the basic diet of vast numbers of people was greatly deficient. The most notable historian of British diet has remarked that, until 1848, we are concerned with a period when 'the diet of the majority of town-dwellers was at best stodgy and monotonous, at worst hopelessly deficient in quality and quantity. Like the agricultural labourer, the town worker in these years survived on the stored-up capital of earlier, healthier generations'.[6]

Whatever improvements may have taken place in the nation's health, there can be no denying the prevalence of widespread killing and disabling diseases and ailments. A host of them swept away vast numbers of people – especially the young – while leaving millions of survivors with life-long crippling and painful infirmities. And the cause and occasion of so many of these afflictions seemed to be firmly rooted in the urban habitat, thus eluding the growing efforts of medical and scientific scrutiny and cure. From birth onwards, the English people of the late eighteenth and early nineteenth centuries faced a daunting series of complaints and physical problems. Moreover, these problems – and their consequences – were more severe, more widespread and more devastating in urban areas. This was particularly so with infant mortality (the death of babies aged 1 and under) which remained high throughout the period (and did not in fact reach its highest recorded level until 1899). From the 1840s onwards, a quarter of all recorded deaths were of babies under 1. Not surprisingly perhaps, infant mortality was highest among the poor, a fact which, in large part, explains the consequent plebeian obsession throughout this period with the need to set money aside for infant funerals. Wherever local research has been conducted, in York, Bath, Preston or Liverpool for example, the evidence is conclusive; babies of the poor died in horrifying numbers. In Preston, in the mid nineteenth century, the mortality rates for infants under 5 among the upper classes was 18 in 100 deaths; among working-class infants it was 62–64 per 100. Not surprisingly, working people were described as 'the industrial or insuring class'. Bad as these figures were, however, they pale by comparison with the death rates among illegitimate children. Of course, the obstacles to infant life were neither new nor restricted to town life. But the evidence clearly shows that the dirt and squalor, which was most acute and most concentrated in urban areas, were responsible (in conjunction with the medical and social customs associated with childbirth) for the appalling statistics.[7]

To claim, however, that there are statistics which are uniquely 'urban' is to mask a markedly varied phenomenon. Within the towns and cities, the statistics of infant deaths (like other contemporary social evidence) varied enormously between different parts of the town. These differences, which were to survive throughout the century and beyond, serve to illustrate a general point; namely that the worst of social conditions determined the worst expectancies of life, and the inescapability of death and illness. That complex but unmistakable skein of social deprivation to be found in the English town and city, of poor diet, foul water, poor ventilation, dangerous sewerage, parental overwork, overcrowding and a general lack of hygiene, confronted the new-born and the young with dangers which were, too often, insurmountable. In 1847, there were, for instance, 75,507 recorded infant deaths in England and Wales. By 1861, it had risen to 106,428, and these figures are clearly *under*-registrations. Furthermore, the loss of infant life was in addition to the startlingly high numbers of mothers who died in childbirth – 3200 recorded in England and Wales in 1847.[8]

It is especially difficult to reconstruct an accurate history of contemporary illness and death if only because of the vagaries of contemporary diagnosis. None the less, the main cause of infant deaths seems clear enough. There were marked numbers of deaths by 'neglect' and 'violence' – though often made worse by current practices of doctors and midwives. Diarrhoea was another major killer of the new-born, and yet, throughout these years, it was regarded as natural and beyond effective medical control. Today, it seems clear enough that the origins of its prevalence lay in unhygienic domestic conditions, food adulteration and the peculiarities of contemporary child-care. Once again, it was a problem which seemed most acute in the cities, and it was not until well after our period that doctors and social reformers were able, through social amelioration, to make significant inroads against this particular scourge of infants.[9]

Other illnesses swept away large numbers of the new-born and the young by tens of thousands. Whooping cough killed some 10,000 annually by the mid nineteenth century – in London it caused one death in thirty. Furthermore, as the century advanced, this was an illness that became more virulent in the towns, but less so in the countryside. While it is true that a child was most at risk in its early months and years, even the relatively 'safe' age group of 4–14 year olds faced dangers normally unfamiliar to the modern western child. Scarlet fever – worse in certain towns and the capital – made deep inroads into this group. When epidemics of scarlet fever struck, they afflicted towns twice as badly as rural communities.[10] Measles, which at times surpassed smallpox as a major killer, was particularly virulent among people who were undernourished. Not surprisingly, it cut swathes through the poor, killing never fewer than 7000 a year throughout the nineteenth century. Typhus – a lice-borne disease – was another killer, hitting the nation hard in periodic epidemics throughout the early and mid nineteenth century. In 1837, 19,000 died; 17,000 in 1847. Typhoid fever was another killer – carried by fouled water, milk or food. As long as the expanding towns remained untouched by modern sanitation and sewerage, this fever continued to afflict the town dweller, especially the poor.[11]

If there is one particular illness which, above all others, characterized, and in many respects obsessed, nineteenth-century society, it was undoubtedly tuberculosis. The figures which plot its devastation among the English people make horrifying reading; its social stigma, like its deathly pallor, have cast a shadow across this country until recent memory. Although the worst ravages of TB declined towards the end of the nineteenth century, it is thought to have killed one person in six throughout these years. Understandably, it was known as 'the white man's disease', as virulent at home as many of the tropical ailments which afflicted Europeans who lived in the tropics. In one year alone, 1838, some 59,000 people died from the disease in England and Wales. By 1851, it caused 3.6 deaths per thousand of the population. Put differently, it has been claimed that TB 'every year killed more people in nineteenth-century Britain than smallpox, scarlet fever, measles, whooping cough and typhus fever put together'.[12] It was a monumental and remorseless killer – particularly among the urban poor, for it colonized the most wretched of urban quarters. The poverty and overcrowding of the expanding, early and mid nineteenth-century city and town formed a fertile habitat for TB. It is a clear reflection of the mounting concern about TB (and other pulmonary disorders) that medical and social observers began to seek an escape from the pollution of urban life in the fresh air of the countryside, the seaside and, later, of the mountains. In fact, the English rush to the seaside and to the open spaces (and, for the prosperous, the establishing of homes on the coast) can only be fully explained by the growing consciousness of the physical harm apparently caused by life in the towns. As with the spa towns at an earlier period, the new seaside resorts were able to benefit by that distinct combination of pleasure seeking and medical advice. Ironically, however, many of the new resorts soon began to reveal many of the same urban problems from which residents or visitors had fled to the coast.[13]

As the nineteenth century advanced and as medical and scientific diagnosis became more 'modern', newer causes of death began to appear. This was especially the case with cancer and heart diseases. Even here early evidence, unsatisfactory as it was, pointed to the social background as the cause and condition of so many ailments. It had been clear for some time that certain ailments were most acute in particular industries. The widely-known lung ailments of miners, respiratory complaints of women in straw-plaiting industries, pulmonary illness of people in the textile trades, the spinal deformities and caved-in chests of people working, from too early an age, in strenuous lifting and carrying jobs – these and countless other examples of illnesses and diseases pointed medical and social attention firmly towards the place of work and the wider environment in which people conducted their daily lives.[14] They were, in fact, particular instances of a general proposition; that a host of contemporary individual and collective illnesses were environmentally-determined.

It would be false to claim that this view of illness and disease, a truism in modern western eyes, was unknown or ignored by contemporary opinion. Nor was it new. What made the matter more pressing – and more difficult to solve – was the very rapid growth of the towns that offered so fertile a breeding ground for this host of ailments. It was becoming ever clearer that medicine (however crude its form) could not make major advances until the urban environment was itself brought more effectively under man's control. And of all the phenomena that served to underline the symbiotic relationship between the urban environment and disease, cholera was pre-eminent. The cholera epidemics of the 1830s and 1840s were devastating in themselves, but had social and political ramifications far beyond the bounds of public health. No other contemporary disease raised such fundamental questions about medical treatment, social policy, and even the fragility of urban political control. It became apparent to more and more informed opinion in the course of those epidemics that the mastery of disease and epidemics ultimately demanded a firmer control over the wider urban environment.

In 1832 and 1849, cholera swept through the country, the first wave coinciding with other social and economic convulsions to create a climate of panic in a number of major towns and cities. It appeared in September 1831 in the new – and pestilential – town of Sunderland; between February–November 1832 it spread rapidly around the country, appearing in 431 towns and villages and afflicting 82,528 people, killing 31,376.[15] From one town to another the attack of the new (for Britain) and incurable disease revealed both the shortcomings of medical science but, more especially, the governmental and bureaucratic shortcomings in the management of towns. The regulations which shaped the local defences against the disease were of seventeenth-century origins, while newer measures tended to highlight the conflict of interests between localities and the agencies of central government in London. The indecision and slowness of the metropolitan response was often compounded by ineptitude and panic in the localities. Yet, the trauma of the cholera, at a time of heightened social tension, helped to convince an enlightened few in London and the provinces of the imperative need for a systematic public health system.

While the cholera was an acute epidemic, its onslaught was against a society not unaccustomed to wholesale death by disease; there were, for instance, 16,000 annual deaths from typhus.[16] Perhaps for that very reason, the panic – and the measures induced by the cholera – passed as swiftly as the disease itself. Many people quickly reverted to their old ways, as a report on health in York observed in 1845:

> On several occasions, when the disease became excessive, and the men in power and the better classes in general became alarmed for their own safety, they gave some attention to the undrained and filthy condition of the localities and abodes of the lower classes, and made some temporary efforts to remove the evils; but that no sooner had the impressive danger passed over than the drainage and cleansing were neglected as before. . . .[17]

Churchmen were, inevitably perhaps, fond of seeing in the cholera the hand of a vengeful deity, visiting his punishments on a sinning people. The alleged sins were varied and sometimes conflicting. Some thought it a penalty for granting Catholic toleration, others for enfranchising dissenters and Jews. One even claimed it was 'to deter people from marrying the sisters of their deceased wives'.[18] The real reason was more prosaic – and was revealed by closer scrutiny of cholera's deathly statistics. Time and again, the poor and the working classes were the major groups to suffer. But the medical and social treatment of the illness was, as with many other complaints, severely hampered by the persistent commitment to the miasma theory of contagious disease, i.e. the belief that diseases were transmitted through the air. The cholera epidemics generated a fierce medical debate, between miasmatists and the contagionists, about the very nature of the transmission of diseases. It was not until 1854 that Dr John Snow, in his book *On the Mode of Communication of Cholera*, convincingly illustrated the true transmission of the disease. In that paper, Snow listed the major long-term measures which would eradicate the disease, notably:

> To effect good and perfect drainage.
> To provide an ample supply of water quite free from contamination with the contents of sewers, cesspools and housedrains. . . .
> To provide model lodging-houses for the vagrant class, and sufficient house room for the poor generally. . . .
> To inculcate habits of personal and domestic cleanliness among the people everywhere.[19]

His premise was simple. 'The measures which are intended to prevent disease should be founded upon a correct knowledge of its cause.' Whatever pain and harm cholera had inflicted on the people, it had also helped to illustrate the appalling (but ultimately manageable) conditions which harboured and stimulated disease in towns.

To the modern historian, the root causes of the worst effects of illness and disease seem clear enough and are easily described. But they were much more confusing to contemporaries. To most they must have seemed a variation of the traditional noisesome and unpleasant characteristics of urban life. Supplies of fresh water and the disposal of sewage, for instance, had rarely posed a major problem in a society (London excepted) of small towns and rural settlements. But, as urbanization grew apace, the towns outstripped the associated facilities of drainage and water supplies. Overcrowded quarters, like London since time immemorial, deposited their excrement into the rivers and becks, or into the ground, through which water filtered into the wells which continued to provide a main source of urban water supplies. In the case of York, a medium-sized town which had none the less increased its population by 20.5 per cent in the 1820s (to 26,260), the bulk of the city's domestic water was drawn from the River Ouse, but downstream of a point where a great deal of domestic and industrial waste was flushed into the river. Similarly, some of the city's wells were close to burial grounds, through which water clearly percolated into the wells.[20] In Gateshead, the population tripling between 1811 and 1851, people were concentrated in the ancient centre of the town; one community of 2000 people had only three privies. There were, in addition, 181 tripe shops, 'numerous piggeries' and thirty-one slaughter houses, the refuse from which was only effectively cleared away by scavenging pigs.[21]

A similar story, with local variations, could be repeated wherever the cholera struck, reminding people, whatever theory of contagion they supported, of the unpleasant nature of urban life. The pressing need for sanitary reform was obvious and ubiquitous, the attacks of epidemic diseases or the publication of appalling health and mortality statistics by investigators and doctors underlining the pressing need for a thorough cleansing of English towns and cities. Yet, and despite the attacks of disease, an air of fatalism hampered the work of reform. Even the 1832 Cholera Prevention Act was prefaced with the remark: 'whereas it has pleased Almighty God to visit the United Kingdom with the plague called the cholera. . .'. There was, however, an abundance of readily available evidence pointing to the links between urban conditions and ill-health. To make English towns healthier it was clear that they needed to be made cleaner, a proposition which was neither new nor original but which was given added urgency by the disaster of the cholera epidemic. And it was a proposition amply confirmed by abundant and increasing data from all corners of the nation. The work of Charles Turner Thackray in Leeds, for instance, illustrated, like others, the need for the supervision and reform of industrial processes. But it was the onslaught of the cholera and the extraordinary work of Edwin Chadwick which provided the most immediate impulse for change.[22]

It was Chadwick's work on the poor laws, infamous in many respects, which fed his methodical mind with the experience and statistics of poverty. At his suggestion the 1837 Act to register births and deaths also included provision to record the cause of death, thus yielding the data for further social scrutiny. In addition, the work of private and government-sponsored researchers into the relationship between poverty and disease (much of it initiated by doctors with extensive experience of town life) helped to accumulate an impressive body of information about,

> the filthy, close and crowded state of the houses and the poisonous condition of the localities in which the greater part of the houses are situated from the total want of drainage and the masses of putrefying matter of all sorts which are allowed to remain and accumulate indefinitely. . . .[23]

The ultimate outcome of this and other investigations, and the presentation of the evidence to Parliament (a process which had been first effectively begun by the campaign against the slave trade), was Chadwick's *Report on the Sanitary Condition of the Labouring Population of Great Britain* (1842). On the strength of this report some local authorities began to initiate their own local reforms, while the government appointed its own commission of inquiry (which Chadwick inevitably greatly influenced). There was also formed the Health of Towns Association in 1844, and, eventually, the Public Health Act of 1848 was passed which has been described as 'the tentative and uncertain start to subsequent government cleansing of the urban environment'. There were, it is true, other parallel drives for sanitary reform (some of which could be traced back to late eighteenth-century roots). Most notable, perhaps, were the local boards of health created between 1831 and 1834, while acts often empowered specific cities to cleanse and reform themselves.[24] We need, however, to be clear that such measures, when effective or implemented, were fully operative only in the very last years of this study. Indeed, the material conditions of urban life and the mortality and health statistics attendant on them, were effectively unimproved until relatively late in the century. If it is possible to characterize the experience of health and illness in these years, it is of a process of widening immiseration due, in large measure, to the rapid spread of urban life, punctuated in the second quarter of the century by the first tentative and inadequate efforts to tackle the problems by initial urban amelioration.

The human problem accentuated by the growth of the towns was as much a political as a medical problem. Indeed, the political urge to combat the social roots of the nation's urban illnesses paralleled the widening medical investigation into the nature and cure of those same ailments. Throughout these years, however, the benefits of formal medicine were, at best, marginal. In the case of hospitals, for instance, in 1800 there were only 4000 hospital beds in England, catering for perhaps 30,000 patients a year (from a population of 9–10 million). Even in 1861, there were only 12,000 beds – for perhaps 120,000 patients annually, but any such analysis is made extremely difficult by the fact that hospitals did not keep full statistics until the 1860s.[25] It is true that most hostpitals had large numbers of outpatients, yet most also had strict restrictions about who (or which ailments) they would or would not treat. Indeed, the admissions policies, the treatment of patients (and equally important, the refusal to treat certain people), the diagnosis and treatment within hospitals thoughout the period is likely to evoke a sense of squeamish outrage in the modern reader. The rapid improvement – and proliferation – of 'modern' hospitals, capable of providing more than a last refuge for the sick and poor, did not materialize until the late nineteenth-century.[26] By and large, hospitals were ineffective and sometimes actually harmful. The evidence recently advanced by some demographers for their successes has failed to note the extremely cautious admissions policies adopted; since hospitals generally refused to admit or treat children, infectious diseases or terminal cases, they stood a reasonable chance of success with those patients who managed to filter through for treatment. Furthermore, the dispensaries and outpatient departments 'were only as useful as the treatment and medicine they supplied, and, in general, these were of little real value'. They may, however, have made one very important contribution; by encouraging among patients a greater personal hygiene.[27]

It would be wrong to overlook the great deal of valiant work undertaken by doctors, particularly in poorer communities. Time and again we know of various medical or local problems and are able to reconstruct medical statistics because of the tireless efforts of local doctors and their assistants. There was, of course, little material reward to be had by those doctors who ventured forth into poor urban communities where medical and urban problems were at their worst. But the doctors were limited, not only by the circumstances of daily life around them, but often also by their own medical or scientific assumptions. Furthermore, it is likely that the very great majority of town dwellers rarely, if ever, encountered a doctor. The medical profession itself was in a state of flux in the early nineteenth century as its ancient organization, like the scientific basis for medicine itself, strove to come to terms with changes in society at large. The old system which divided medical practitioners into three groups – physicians, surgeons and apothecaries – was clearly as inadequate as the medicines they prescribed. And it was the evolution of the general practitioner – whose work cut across traditional professional barriers – which came to provide greater medical care for ever more people. It was, notably, the growth of a prosperous middle class which provided the human and financial basis for this medical change.[28] In the course of the nineteenth century there was a marked 'professionalization' of all branches of medicine – as with other areas of English middle-class life. But these changes really belong to a later period; throughout these years, the medical profession generally lay beyond the ken or the pockets of the great bulk of urban people, though local apothecaries (themselves subject to increasing training and regulation) were widely available for consultation and the prescription of drugs and medicines.

Doctors were often deeply distrusted and disliked, especially lower down the social scale and notably after the grisly revelations of the body-snatching careers of Burke and Hare in 1827–8. Doctors, after all, had a legitimate need for an unending supply of cadavers to satisfy the expanding medical schools and the growing demands for dissection among ever more doctors. But, long before the outcry against body-snatching, doctors had been disliked. By and large they were the source and occasion of great suffering and were often able only to watch over the last painful days of the sick person. Their ability to ease the patient's pain – to say nothing of their ability to cure the ailment – was strictly limited. Often they seemed able only to inflict more pain and suffering – and often to no avail. Time and again, particularly at times of social tension, doctors and local 'quacks' found themselves abused and attacked. When, in September 1832, a body-snatching riot broke out in Manchester, 'The cries of execration against the doctors was very general and an almost unanimous shout of "To the hospital, pull it to the ground" took place.' Moreover, this antipathy to the medical profession, widely manifested throughout the towns of England in the 1820s and 1830s was also common throughout western Europe at much the same time and for similar reasons.[29]

Among the very poor, the doctors were often thought to bring, not an easing of pain and suffering, but the threat of death and even of body-snatching. Clearly, such fears were greatly exaggerated, and were not common to all levels of society. None the less, at those important crises of early nineteenth-century England where epidemic attacks fused with economic and political unrest to produce spasms of discontent and collective violence, we are able to catch a glimpse of the popular mentality about formal medicine. Those doctors who, in 1831–2 and 1849, bore the brunt of the fight against the cholera found themselves distrusted by powerful men in Whitehall (notably Chadwick), ill-considered (and underpaid) by the local poor law guardians and actively disliked by wide sectors of the lower orders. It has been argued that, at certain times, the popular distrust of doctors developed into a specific form of class consciousness.[30] For our purposes, however, it is much more useful as an indicator – inadequate and fragmentary as it was – of popular distrust of doctors and the widespread resentment displayed towards medicine itself. Whatever benefits medicine gave to the urban poor, it was not reciprocated by confidence or support. This goes some way to illustrate the very limited expectations held of doctors and their craft. Whatever evidence some historians have found for the improving results of medicine among the urban poor, they were results which were simply not appreciated by those who ought to have regarded themselves as the beneficiaries of medical efforts.

The role of the early nineteenth-century urban doctor was unenviable.

'They aspired to be regarded as scientific yet critics frequently dismissed them as humbugs. Doctors wanted to be accepted as gentlemen when the great majority knew themselves to be parvenus.' And, after the 1834 Poor Law Amendment Act, they found their work among the poor both more taxing and less rewarded.[31] Of course, within the workhouse, the poor (especially the old) were unavoidably subject to the diagnosis and treatment of contemporary medicine. This was, however, one of the few areas where doctors and poor met with any degree of consistency or regularity. Many more people, however, were obliged (or chose) to rely upon the customs and potions of popular folk medicine handed down since time immemorial. Indeed, it seems clear that folk medicine was a major strand of pre-modern culture that survived in a modernizing world. It would, of course, be absurd to denigrate folk medicine as unscientific and ineffective. After all, it ranged from the oddest and sometimes most counterproductive of superstitions and practices through to highly effective diagnosis and treatment, though using the natural substances which had, since time immemorial, formed the basis of medicine itself. There were myriads of rituals which were thought to stave off or cure illnesses; ceremonies involving donkeys – with patients paraded around the beast – customs involving 'magical' numbers, the use of frogs or mice.[32] Potions, key words, significant people in the locality trusted to carry out these tasks, peripatetic 'quacks' and herbalists – these in addition to the private, family folklore of diagnosis and practice, which passed from one generation to another, providing people (in town and country) with their sole form of medicine. It seems likely throughout these years that ancient, informal, and well-tried practices were more widely-used than the services of professional medicine. Of course, there were places where the two schools overlapped, where folk and 'scientific' medicines were at one both in diagnosis and treatment. But it is an indication of the encroaching professionalization of medicine that more and more of the old practices were declared illegal, removed from lay control and placed under the formal aegis of institutional medicine. This happened in every field of medicine – midwifery, surgery, or pharmacy, for instance – as newly-organized professionals drove a wedge (in the form of restrictive practices or even legislation) between their own work and lay practices.

This change in the practice of medicine was notable in the provision – and restriction – of drugs. Until 1868, the sale of drugs was largely unrestricted. Opium, along with alcohol, was the most common and widely-used form of pain-killer, and was, like any other commercial item, freely available in local shops. Shopkeepers concocted their own opium-based recipes in pills or liquids to deal with a range of ailments. Understandably, this drug 'was popularly used for a whole range of conditions for which medicinal aid would not be generally available'. In the words of the most recent historian of this drug, opium was 'the aspirin of its time' used to sedate fretful or sick babies and children, while offering, incidentally, a useful palliative for the gastro-intestinal ailments so common to urban life and which so afflicted the young.[33] The available data on the sale of those opiates most commonly used for infants suggests staggering and widespread levels of drug consumption. 'Godfrey's Cordial', an extremely popular liquid opiate used for calming troubled children, had massive sales throughout the country. It was calculated that ten gallons a week were sold in Coventry – enough for 12,000 doses. Elsewhere, one man confessed to selling 200 lbs of opium annually. Observers – and doctors – denounced the widespread commitment to opiates among the poor, yet formal medicine was no less committed (for the same reasons) to the adminstration of opiates, as pain-killers, pacifiers and cures, for a range of complaints. In fact, the medical and political attack on the open and commercial use of opiates in the nineteenth century can readily be construed as yet another attempt by respectable society to undermine the cultural practices of the common people, while, none the less, persisting in their own particular commitment to the same drugs.[34] As long as poorer people were permitted (or could afford) to, they turned to the cheaply available remedies of opiates when formal medicine was unobtainable or useless, and this was overwhelmingly true for the commonplace problems that were occasioned or worsened by the social conditions around them. It is, then, not surprising that the use of opiate drugs (or alcohol) apparently increased throughout these years, especially among the lower orders. Many of their ailments, notably the pulmonary and gastro-enteric, were especially prevalent or acute in the urban habitat. As ever more people found themselves caught up in urban life, they were, unavoidably, exposed to the particular disease environment of town life. Opiates might – or might not – provide a cure, but they could, in many cases, offer an easing of pain and discomfort. The better-off could escape, temporarily or permanently, to parts of the country that were safer and healthier. Working people were rooted – in ever-increasing numbers – in that complex web of social deprivation which bred so much ill-health. Until banned in the 1860s, opium, not religion, was the pacifier of plebeian pain and distress.

It would be untrue, however, to claim that opium alone catered for the physical or psychological troubles of the English people. There were, as we have seen, a host of 'natural' products, herbs, roots, etc. which were culled from nature itself. But there was also a range of patent medicines which were bought in profusion by the English people. James Morrison, who claimed that his pills were good for 'fevers, measles, scarletina, smallpox, consumption, lassitude and debility from old age' sold, in 1834, (and not surprisingly, in the light of his claim) more than one million boxes of the pills. Often, of course, people turned to patent medicine when formal medicine had failed them. And it is some indication of the shortcomings and inadequacies of formal medicine that patent medicines and pills were sold by the millions of doses, their inventors able to become millionaires by capitalizing on people's hopes or desperation. It is true that one extreme type of the patent medicine manufacturer was a charlatan, passing off pills made of wax, lard or turpentine, but, at the other extreme, were the pioneer herbalists and chemists who created medical empires which survive to this day; Thomas Beecham and Jesse Boot are obvious examples.[35]

Just as patent and folk medicine supplemented formal medicine, there were also legions of quacks ready and willing to administer to the sick, the anxious and the worried – and the gullible. Self-proclaimed doctors, travelling 'professors' and surgeons – with no training, but an abundance of self-confidence and a persuasive line in salesmanship – travelled the country selling their curative skills and potions. Inevitably, since they appealed to desperate people, often in the last stages of painful illnesses, they often did little but supervise the death throes, in return for a fee. One man who established a considerable reputation for himself in Manchester in the 1840s was, symbolically perhaps, Dr Coffin. Such quacks (not all as bad or as criminal as we might imagine) were but one type of a larger species of informal practitioners of the curative arts embracing the traditional folk-healer – be they herbalists or 'witches'. While, perhaps, the bulk of their trade came from poor working people, they were also able to secure business among more prosperous people. But it was among the poor – who often saw in formal medicine little but pain and indifference – that the 'quacks' were able to make such lucrative (if medically ineffective) inroads.[36] Indeed, we can detect in the widespread commitment to folk medicine and quackery the hostile popular mentality to the practices and failings of formal medicine itself. By mid century, the people of urban England had few benefits to boast of from the encroaching world of formal medicine. Indeed, what marginal benefits had come their way had resulted from the early and almost indefinable combination of economic amelioration and urban improvement – changes which were, however, to transform urban life late in the century.

The health of this increasingly urban nation was clearly a monumental problem. And, bizarre as it may seem, few of the problems of health seemed worse, more grotesque, and dangerous than the difficulty of disposing of the dead. Yet, few historians have attempted to treat the problem of death and dying in these years as one of the most taxing and sensitive issues troubling the late eighteenth- and early nineteenth-century English town and cities. Burying the increasing numbers of bodies was, at once, a medical and a social problem which, like the wider issue of illness and the urban environment, demanded private, municipal, and, ultimately, national intervention. Indeed, one could argue that, of all the problems facing the urbanizing society, few were more unpleasant and so complex a matter of hygiene and religion, and so elusive of political solution, as the disposal of the dead. In large measure, the problem had been mastered by the second half of the nineteenth century but, even then, between 1852 and 1899 some forty-five Acts of Parliament were passed regulating the various problems of burial.[37]

By the early years of the nineteenth century it was obvious, in one city after another, that the old burial grounds, especially the churchyards, were, like so many other urban facilities, utterly inadequate to meet the new demands placed on them. And, just as the towns themselves expanded, so, too, did the numbers of burial grounds. In the thirty years to 1863, some 3000 new burial grounds were opened to cope with the Victorian dead.[38] But, in the first thirty years of the century (and, in some areas, long after) burial grounds were filled to overflowing, creating scandalous, distressing and harmful scenes. Soil was imported and dumped on top of the burial grounds to provide a few more inches of burial space; 'every foot of the ground has been occupied and re-occupied and re-employed – until the very soil, compounded as it is of human remains, constitutes, in the warm season more especially, vast hot-beds of pestilential infection'. The stench from the urban burial grounds and the apparent ill-health of those living nearby, or, worse still, who worked there, seemed to strengthen the miasmatists' theory. 'Men accustomed to the work of grave digging have been suddenly struck down by concentrated emanations from the dead.'[39] London offered the worst examples and it was there that medical opinion first began to demand reforms in urban burial and the development of burial grounds *away* from the centres of population. According to one London doctor, burial grounds within the towns were 'a national evil – the harbingers, if not the originators, of pestilence; the cause, direct or indirect, of inhumanity, immorality, and irreligion'.[40] Yet, London presented only the most obvious and extreme example. York, for instance, had precisely the same problem. There, it was claimed in 1847,

Corpses are consigned to our crowded churchyards, already sodden with human flesh and gore, or deposited in a damp and ill-closed vault, either under the pew of some family in the parish church, or in the aisle, where probably a stove is erected, the heat from which draws forth the noxious fumes of corruption and the unsuspecting congregation inhale that, which, in many cases, has terminated in death.[41]

It was impossible to find space for further burials in twenty-four of York's churchyards, and even then, some coffins had been deposited only inches from the surface – but only after extra soil had been dumped on top. Tales of dogs waiting for human scraps and of children playing with skulls were commonplace and plausible.[42] By mid century, the local council and visitations of clergymen were adamant that none of the parish churches within the city were fit to receive more bodies, since all were 'overcrowded with human remains'.[43] It was to overcome these distasteful problems that, in common with other towns, the York Public Cemetery Company was founded in 1836–7 on an 8 acre site outside the city. Some indication of the massive population growth in the second half of the century can be gauged by the fact that, by 1904, it extended to 13 acres and was half full. By 1958, almost all of its 50 acres were filled with graves.[44]

It was characteristic of the new municipal ideology in the early nineteenth century that the initial solution arrived at was the creation of cemetery companies. The business and financial institutions which had already proved so successful and instrumental in creating the first steps towards an industrial society were thought to be proper for the disposal of the dead. At first, commercial cemeteries seemed to make commercial sense. The St James' cemetery in London paid a dividend of 8 per cent in 1839. Between 1832 and 1847, Parliament authorized eight new commercial cemeteries for the capital. Yet, the move towards new commercial cemeteries, designed with health and aesthetics in mind, did not begin in England until 1824. Thereafter, however, some cemeteries established themselves as major landmarks in early and mid Victorian urban architecture, landscaping and design, although most were initially sited outside the city, in neighbouring rural positions – Highgate and Kensal Green, for instance. It may seem ironic that some of the first effective steps towards modern urban planning – the creation of a totally manmade habitat pleasing to the eye and socially useful – were taken for the dead, and only indirectly for the living. Indeed, the Victorian cemeteries became (and have, in many respects, remained) major objects of admiration. When, in 1849, Victoria and Albert visited Glasgow, they visited and admired the Glasgow Necropolis.[45]

Within the new – and private – burial grounds, there was created a new urban environment; not merely a place to bury the dead, but a landscaped and regulated parkland that people could visit and, after a fashion, enjoy as they might a visit to a park. Such cemeteries, in the words of one designer in 1830, 'ought to be made sufficiently large to serve at the same time as breathing spaces'.[46] By the mid 1830s, cemeteries attracted the skills of the best of contemporary designers, architects and engineers, their finished products thought to provide a pleasing setting which 'could only be beneficial to public morals and manners'. There were, by mid century, even specialist journals, manuals and books on the design of cemeteries.[47]

By and large, however, the new commercial cemeteries had few attractions for the lower orders – for obvious reasons. While these developments were directed at more prosperous families, perhaps the most pressing need was to provide a more decent burial for the poor and working people. Because of low incomes and the dreadful mortality rate among their infants, poorer groups had long been committed to saving for the unavoidable expenses of burials. Throughout the eighteenth century, Friendly Societies had been a feature of rural England – encouraged by local men of property anxious to avoid having to pay for paupers' funerals. But it was in the course of the nineteenth century that Friendly Societies blossomed, with their greatest concentration of members in the industrial north-west; upwards of 5 million by 1875.[48] What gave the Friendly Society an added filip was the 1834 Poor Law Act and its repercussions upon the time-honoured customs of dying and burials among the poor. Death, after all, was a domestic experience; people died within the home and the corpse was left in the home until the funeral. Of course, this often created a series of health hazards.

> The sides of a wooden coffin, often imperfectly made, are at best all that divides the decomposition of the dead, from the respiration of the living.[49]

Throughout plebeian life, it was common for neighbours, friends, workmates, and relatives, to gather round the coffin to pay their last respects and to enjoy the particular Wake dictated by local custom. It was all medical men could do to stop the bereaved touching and kissing the corpse, even when it was diseased or putrefying. Moreover, a corpse might lie in the home for more than a week – until the family was able to save or borrow enough to pay for the funeral. It was understandable, then, that when, from the 1840s onwards, new cemeteries opened for business, they generally provided a mortuary – part of the growing determination to remove the corpse from the home.[50]

For the very poor – the paupers – the 1834 Act had already achieved the same result. Paupers died in the workhouse, away from that web of family and social custom which had marked their passing since time immemorial. Additionally, the ignominy of a pauper's death was compounded by the shame of a pauper's grave. Private, parish, or municipal burial grounds set aside remote patches for pauper burials, to be paid for by the local Poor Law Guardians. But the paupers' were anonymous and communal, with no headstone or marker to signify who – or how many – were buried in that spot. Thus, the burial, like dying in the workhouse, allowed no room for those traditional acts of mourning and remembrance which, at all levels of society, formed such cherished and significant ceremonies. Nor was this merely the result of accidental change, for among those who adminstered the poor law, there was a clear determination to drive out the traditional celebration of death among the very poor. Observers had long been bemused that the poor were prepared to lavish such time and expense on bereavement. In 1843, Chadwick noted

> the desire to secure respectful interment of themselves and their relatives is perhaps the strongest and most widely-diffused feeling among the labouring classes of the population.[51]

In order to stop this, large numbers of Poor Law Guardians refused assistance to 'the widows of men who have been provident enough to insure for the funeral expenses, if, in the opinion of the Board of Guardians, such funeral money had been "lavished or improperly expended" '. Some indication of the anonymity with which Guardians shrouded the burial of the paupers can be seen in the failure of those persistent researchers, the Webbs, to find scarcely a mention of pauper burials in the 'vast stream of official reports on poor law administration'.[52] Those who died in the workhouse effectively disappeared. Their final resting place went unmarked and their remains denied even the cold comfort of a mention in the records. Not surprisingly, those who could strove to avoid this fate. In 1843, Chadwick caculated that, of £24 million in savings banks, between £6–8 million was specifically set aside for funeral expenses.[53]

Ironically, the amelioration in burial grounds increased the cost of funerals. They were, after all, outside the towns and necessarily involved a journey, which often turned into a procession for those people belonging to institutions – a Sunday school, ragged school, a union or a choir. And it was a more costly journey even for the poor; the use of the handcart or cab to transport the coffin cost more the further the distance travelled. It is equally significant that the new cemeteries charged differential rates not merely for different-sized plots and monuments but also graded their charges according to social class. In Tiverton, for example, the cost of the hand hearse 'for an adult labourer was 2/6d. The like, for a tradesman or any other person, 5/-'. Victorian burial regulations often read like a litany of every conceivable gradation and condition of contemporary society. By mid century, Victorian society had come to marshal and to describe the dead just as it tabulated and perceived the living; in the rank order of social class.[54]

The new private or municipal cemeteries which, by mid century, fringed English towns and cities provided more than a mere resting place for the dead (there were, for instance, half a million burials in London in the 1840s).[55] They were also landmarks on that rural borderland which was readily accessible to most urban dwellers. In London, the cemeteries at Kensall Green, Norwood, Highgate, Nunhead, Tower Hamlets, and Brompton, were all in a rural setting. It is some reflection of the continuing – and indeed accelerating – process of urban expansion that these and others throughout the country were engulfed in the second half of the century by the spilling outwards of the towns and cities. Yet, the mid century cemeteries are important reminders (and, of course, survivors) of some of the most potent social forces which created the nature, quality, (and problems) of early Victorian life.

Throughout much of this period, the English people continued to see and experience dying and death close at hand and normally within their own house. Inevitably, if the home itself was overcrowded and squalid, so, too, would be the deaths in that family. In more prosperous, bigger homes, the terminally sick could be isolated in their own room, but this was clearly impossible for the very great bulk of town dwellers who, as we have seen, lived in overcrowded domestic circumstances. It might be felt, however, that there was nothing new in this. Yet we need, once again, to return to the overall changes in demography and living patterns. As ever more people lived in towns, the bulk of them living a materially meagre life in cramped domestic circumstances, the problems of dying and death were consequently multiplied. The obvious reflection of this is, as we have seen, in the serious problems of burial and amelioration of the urban burial grounds.

There is, however, a wider, though less tangible historical problem involved, for throughout these years contemporary English people were exposed to social experiences which would be quite alien to their modern descendants. Today (in the west at least) pain and dying are, by and large, controlled and institutionalized, and it is difficult to imagine the effects of seeing and enduring suffering within the home – and at a time when so little could be done to alleviate it. It would be quite wrong, however, to imagine that familiarity with such problems hardened people towards them. It is naturally difficult to explain the full ramifications of the awareness and proximity of suffering and death on the complex make-up of individuals and of society at large. Recent studies of autobiographies clearly show the profound psychological (and religious) consequences of the unexpected, sudden, often wholesale and close-at-hand death of loved ones. The working-class radical, Samuel Bamford, lost his brother, sister, grandfather, uncle and mother in an epidemic:

> What a void around us, what a diminished and unsheltered group we seemed to be! Surely 'the bitterness of death' is in the lonesome desolation of the living; and this bitterness, notwithstanding my naturally cheerful temper, and all which kindness could do to console me, was long my portion, until I feared whether or not I should ever be called from 'the valley of the shadow of eternity'.[56]

As we have seen, the major killing or disabling ailments in English society were rooted in the unwholesome conditions of urban life. Whereas today modern western man can expect a wide range of urban facilities to come to his aid, few English people in the years up to the mid nineteenth century could realistically expect or hope for such good fortune. Indeed, it seemed for much of this period that the English were becoming an unhealthier people in proportion to their urbanization. As ever more English people found themselves living in towns or cities, life continued to be both as nasty and brutish – if not quite as short – as it always had been.

3 *Working lives*

The most formative and seminal force in the lives of the great bulk of the English people was work. This had been true, of course, since time immemorial and was to remain so long after our period. From the tenderest of years until old age, infirmity or death intervened, labouring people were predestined to do just that – labour. For many of them, however, these years witnessed a qualitative shift in their working lives – and not for the better – as they found their lifetime's labours located in an urban environment. Moreover, in many of the nation's expansive industries, the very nature of employment was of a qualitatively new kind, for which no amount of previous experience – of long hours, application and endurance – was of particular benefit or relevance. An increasing number of working people were, at one and the same time, expected to adapt their domestic lives to the rigours of the urban environment and the rhythms of their working lives to the demands of new industrial processes.

Among the most popular and durable images of plebeian life in the nineteenth century are pictures, real and imaginary, of toil in the new industrial enterprises. Bleak impressions, especially of child and female labour in the new factories, workshops and mines have in large measure become part of the national folklore handed down from one generation to another. And yet it is not to deny the existence or the severity of appalling labouring experiences to suggest that they ought to be placed in a wider context of work in these years. To state the matter crudely, the unpleasantness of child and female labour – or even the worst forms of labouring exploitation – were not always a function of the new industrial processes, nor were they restricted to the newer industrial systems.

First, however, it is crucial to establish a distinction (one which is often blurred in the public mind) between urbanization and industrialization. It is so often assumed that these two phenomena are either one and the same or are necessarily linked. Now, while it is undoubtedly true that many embryonic industries developed and thrived in growing urban settings, substantial industrial development took place in a more rural environment. This was strikingly so in the case of mining where, to this day, mining communities are often 'mining villages'; heavy and modernized industrial activity concentrated in an overwhelmingly agricultural environment. Similarly, some of the worst human suffering, at a personal or collective level, was a result not so much of industrial as of urban life *tout court*. This is not, of course, to deny the particularly debilitating afflictions caused by certain kinds of work or the industrial environment – be it in mining, textiles or pottery. We need, even here, to recall that industrial and occupational ailments were also commonplace in rural settings; pulmonary complaints among girls working in cottage industries, rheumatism among labourers exposed to the cold, rain and wind for long periods, returning to damp and cold homes and then returning to work in damp clothing. But what so often made many of the expansive towns so unpleasant – so uncomfortable and even dangerous for new generations of plebeian workers – was the close proximity of work and home. As we have already seen, home was generally a squalid affair. Making it worse was its closeness not merely to overcrowded neighbours but also to local industries. Town dwellers, quite simply, lived within walking distance of the noise, the smell and the lights of the workplace, and this was a major factor in rendering their home lives even more restless. However mean the agricultural labourer's home it was at least usually in a more peaceful setting. For growing numbers of people a good night's sleep was hard to come by, a fact that was to have detrimental effects on the physical growth of the young whose sleep, so vital for their growth, was often disrupted by the noises from the homes, industries and the town itself. Few places could generate such a tumultuous noise as an early factory in full operation. Despite this, it is important to place the factory in its proper historical and social setting.

The shadow of the factory

Perhaps the most popular and widespread of all the images of England in these years is the dominance of the new factory system; of regarding English life and labour ever more dominated by modern production systems and determined to an increasing degree by the encroaching influence of certain key industries. The factory soon came to symbolize and to represent the characteristic workplace of English working life. It was one of those new English institutions visited and commented on by successive waves of foreign visitors. Yet the factory was, in many respects, unusual and untypical – except in certain industries and regions – and throughout this period the factory tended to be a small-scale operation.

There was, it is true, a rapid development of the factory system, harnessing and utilizing the advances in technical changes especially in the years after the French wars. In textiles, in particular, the factory came to predominate and, as it did so, the products of factory labour began to spread out, across the nation and overseas, to transform the clothing of millions and to enrich the nation at large. By 1851 more than half a million workers were employed in the cotton industry; in 1841 83 per cent of them worked in factories, whereas only 44 per cent did so in the woollen trades. There could be no denying the vast increase in annual output generated by the shift to a factory system in key industries – and of the national economic benefit. By 1830 50 per cent of the nation's exports were cotton products.[1]

Cotton was the unquestioned pacemaker of industrial change, utilizing and then stimulating further technical change. Cotton relied upon the enslaved labour first of the Caribbean and then the US South, and the factories were working by a rapidly expanding labour force – much of it women and children – whose working and social lives were, in this period of flux, commonly believed to be greatly worsened. The technological changes behind the cotton revolution were, until the mid nineteenth century, relatively simple and although the processes were helped by steam power, it was not always essential. As late as 1838 for instance, water provided one-quarter of the industry's energy. But the process greatly accelerated after 1815: power looms in England increased from 2400 in 1813 to 224,000 in 1850.[2]

This fairly rapid transition from water to steam power is symptomatic and we ought not to lose sight of the fact that the new industrial processes evolved from a combination of the novel and well-established. Moreover, this process of transmutation was also at work among those labouring people involved in textiles; just as mechanical processes had to adapt, so too did the people tending to them. Most crucially, they had to adapt to living in towns; those emergent urban areas which grew like an unkempt, and initially uncontrolled dirty cocoon around the industrial workplace.

One obvious and visible consequence of the development of the factory system in Lancashire and Yorkshire was the proliferation of the factory towns – a particular form of the urban process already well in train *before* the effective emergence of the factory system itself. From contemporary accounts of those distinctive northern towns and their complex of peculiar social problems, there was established early on a view of the industrial revolution which remains so widely-held today. It is abundantly clear that these new factory towns lured people from the immediate rural environs to seek work in the expanding textile trades. Some measure of this migration of labour from the country to the town has been gauged by researches into the particular demography of certain towns. In 1851 70 per cent of the adult population of Preston had been born outside the town, although the proportion was significantly lower (50 per cent) in the older textile centres of Bolton, Manchester and Stockport.[3]

When people moved into the textile districts it did not, normally, involve quite the leap in the dark so commonly imagined. People tended to move short distances, though often moving on yet again in a 'leap-frogging' process of urban migration. In Preston, for instance, almost one-quarter of the total population had been born within 10 miles of the town, and many of those had already worked in rural industrial undertakings. Nor did this population movement disrupt ties of kin and culture since migrants to the towns were often given an initial base and *entrée* to town life by relatives and friends already established there.[4] This is, after all, true of those foreign immigrants who settled in England throughout the eighteenth, nineteenth and twentieth centuries. In these years it was particularly true of Irish immigration. Immigrants moved into the same towns, the same districts, the same streets and even the same buildings as relatives, friends or fellow villagers already established. This was true of the eighteenth-century Irish (and twentieth-century Jamaicans) and it ought not to surprise us that a similar pattern evolved among rural folk settling in the towns. There has, however, been a temptation, to which large numbers of historians have succumbed, to view the factory towns of Lancashire and Yorkshire as the crucibles into which were poured the human ingredients of a new industrial and urban life form, and from which there poured forth a subdued and de-racinated labour force. Among other things, such a view ignores the remarkable persistence and survival of traditional links, mentalities and values. By the mid nineteenth century, the dominant social commitment of working people in northern factory towns seems to have been to kin, and thence to the immediate neighbouring community. It is true that within that community the factory was clearly of paramount economic importance; 'the factory and its owner . . . were central elements in the forging of the neighbour bond . . .'.[5] This 'sense of place', so central to the textile communities, was shaped by the place of work and the structure of authority that flowed from it. None the less it seems clear that in the factory towns (though not in the bigger cities) the nature and quality of community life was much more that of a village than of a city. And this sense of immediate community, created in the shadow of the factory, was consolidated from mid century by the general reduction of the flow of population from the countryside into the towns. Working people in the textile trades belonged to closely-knit networks of kinship and community which, though forged by the immediate dictates of work and authority, were suffused with values and qualities derived from rural life (and that rural life was close in both distance and spirit).

We need, then, to be extremely cautious in assessing the extent and nature of the impact of factory life in these years. It is important to stress just how limited, by mid century, was the process of bringing manufacture under a modern factory roof. The actual *size* of factory operations was small. By mid century the average spinning firm employed 100 people, weaving 108. In 1851 in textiles, only 411 employers, out of 1670, employed a workforce larger than 100 people. Much the same was true of engineering companies; four out of five employers had fewer than ten workmen. Robert Owen's New Lanark Mills employing 1500 in 1816 were quite exceptional.[6]

In the textile towns, working people lived overwhelmingly within a short walking distance of their place of work; able to be woken up by a factory hooter or a factory employee. Many lived, quite literally, within the shadow of the factory. It was much later in the century that new urban transport systems (plus the consumer power to travel on them) enabled working people to distance themselves from the place of work – and from the pervasive influence of the wider culture of the workplace and the influence of the owners. Much later still, the further expansion and modernization of the textile industries, continuing urban growth and the widening geographic and political distance between, on the one hand, ownership and management and labour on the other, exacerbated the social tensions of late nineteenth-century urban life.

The ramifications of the textile (especially cotton) trades upon the economic life of the nation, and indeed upon a great deal of northern urban development, is incalcuable. Textiles clearly generated remarkable urban growth in their region, creating a cluster of satellite towns held in the gravitational pull of the major economic centres of Manchester and Leeds. The consequences, however, were felt even further afield. Liverpool, the entrepôt for US (later Indian) cotton, developed its own economic structure of brokers, merchants, and workers. There also emerged 'the developing nexus of communications along the Mersey valley'. While it is true that a host of academic critics in the 1960s and 1970s have challenged the orthodox view of cotton as the pacemaker of industrial change and growth, it is hard to deny its centrality in a number of ways. However, its importance greatly depends on the criteria we choose for assessment. Throughout the first fifty years of the nineteenth century cotton alone employed between 4 and 6 per cent of the total labour force of the UK. In Lancashire over one-third of working people were employed in cotton.[7]

The expansion of the physical plant attendant on cotton's development was remarkable. The industry called into being 'mills, machinery, power, light and mill stores'. This in turn created demand for 'bricks, slate, timber, glass and iron' – with all the inevitable consequences on industrial development in those occupations. And of course, the need for power as the industry became increasingly mechanized created a related growth in coal, particularly in the Lancashire coalfield. One product after another was devised, transformed or enhanced by the voracious expansion of the textile trades.

One of the most obvious direct consequences of textile expansion was the consumer power of the textile operatives (substantial numbers of whom were female) with a resulting impact on personal and domestic goods and services. This must, however, be set against the decline and immiseration of the displaced handloom weavers.[8] There were a quarter of a million handloom weavers in the 1820s, 100,000 by the early 1840s, and only 50,000, in the most wretched of conditions, a decade later.[9] When we look at their miserable story there is little evidence in their experience for the widening diffusion of material benefits generated by industrial change; they were the most obvious (though not the only) casualties of economic change. But even this cautionary qualification cannot disguise the massive expansion of consumer power in towns and cities throughout the country. Even before the coming of the new industries the eighteenth-century towns had established themselves as major acquisitive and consumer societies, devouring the goods, the luxuries and the essentials, the basics and the fripperies disgorged in growing volume by both rural and urban industries.[10] This was, however, greatly accentuated by the encroachment of new industries and new industrial towns. Furthermore, it was a consumer power that was satisfied by marketing and retailing systems which in many respects bore a close affinity to the traditional pre-industrial systems. Local neighbourhood shops (general or specialized), skilled craftmen who both produced and sold their wares (tailors and shoemakers for instance), regular (normally weekly) markets for agricultural produce were all, together with many others more than mere extensions of an old and traditional system. Even in the newer industrial towns markets quickly grew up, resembling in almost every respect the ancient marketing tradition of the old county towns. Completing these retailing services were bands of itinerant traders, salesmen and hawkers peddling their wares from town to town and from street to street. And in most of these occupations craft skills, or the art of salesmanship and marketing, were considerable and were often handed down within families from one generation to another.

It might be felt that since the great bulk of the labouring population, in town and country, endured poor housing, miserable diet and sparse clothing, the condition of labouring people provided little scope for advancement in the retail trades. There were, however, two factors which stimulated those trades despite the unquestioned hardships of the majority of plebeian consumers. First, and obviously, there was the sheer rise of population, and second, more and more people, notably the middle class, were able to enjoy more than the mere necessities of life. For the middle classes there developed a distinctive town trade of grocers' and retailers' shops, whereas plebeian consumer demand tended to be satisfied by the markets, fairs, pedlars and travelling salesmen. By the mid nineteenth century, despite certain obvious changes in retailing and distribution, particularly the early impact of the railway system which was able to create and satisfy a national consumer market, the system was recognizably that described by Daniel Defoe, for instance, in the early eighteenth century. A historian of the retail trade has described it thus: the 'small scale of the units engaged in the trade, and emphasis on skill and experience in retailing, the haggling as to price, and the important role played by open markets, had not changed fundamentally'.[11] All this, however, was to change in the second half of the nineteenth century.

For our purpose what is striking is how traditional the system of feeding and clothing the English people was. Here, once more, we find an essentially pre-industrial system surviving and infusing itself into the urbanizing and industrializing world of the early nineteenth century. There is, again, unmistakable evidence that the new social order which served the urban world did not so much replace or push aside the older, traditional one, but, like the towns themselves, simply grew on top of it. This is then a specific example of a general argument that runs throughout this book and which provides its intellectual core, namely that in a society undergoing the most fundamental of human and social changes, the survivals and continuities with older structures and customs remain among England's most durable and formative characteristics.

Even by the mid nineteenth century, however much the English might pride themselves on their industrial pre-eminence and achievements, there were vast areas of their economy, population and culture which appeared to be only relatively affected by the urban and industrial transformations. In essence this is a continuing reminder against the commonplace impression that industrial progress drove all before it, transforming England into a 'modern' industrial society which was rapidly breaking its links with the pre-industrial past.

This becomes clearer when we take a closer look at work itself. Despite the undeniable advance of more modern techniques, a great deal of industrial employment took traditional forms. Outworkers remained vital in a range of industries. It is of course true that in the north of England where textiles predominated the outworkers were progressively reduced, by stages, from primary industrial agents, to ancilliary roles until finally reaching the nadir of their fortunes as depressed and disappearing remanants of a past society. By the later nineteeth century, where domestic outworkers survived in the north they did so 'chiefly in the mass-production clothing trades of the great provincial centres – Leeds, Manchester, Sheffield and Newcastle'.[12] Both in textiles and heavy industries outworkers had been reduced and then eliminated, as the manufacturing processes were increasingly mechanized and brought inside the factory or plant. But this process was not as pronounced in other industries, with the result that by mid century many industries continued to be characterized by and dependent on large numbers of domestic outworkers. For tens of thousands of outworkers, then, it is quite impossible to draw a distinguishing line between home and the workplace. Many people grew up and lived in homes which were also the centre of economic activity. They experienced the age-old disadvantages of work taking place in the home in conjunction with the distinct deprivations of urban life.

Large numbers of key industries followed the transformation in textiles; potteries, brewing, engineering, shipbuilding, metal, glass and chemicals all turned increasingly to larger scale factory-based productions, with many of the human and urban consequences already seen in textiles. Yet small-scale production remained widespread throughout mid-century England. In the Midlands a vast array of metal goods, 'weapons, locks, screws, nails' and more besides, were manufactured in small workshops: In the nail trade, for instance, the high-point of the outwork manufacturing system was *c.* 1830 when some 50,000 people were employed throughout the region; there were as many as 10,000 as late as the 1870s. In certain industries outworking actually increasing towards the end of our period, notably in the case of chain-making. A similar pattern developed in the footwear, hosiery, silk, lace and straw industries in the East Midlands.[13] And much the same was true in London where, in the words of Gareth Stedman Jones, 'The effect of the Industrial Revolution . . . was to accentuate its pre-industrial characteristics.' Indeed, large-scale factory operations in the capital were actually rare, for a variety of largely economic reasons. Moreover, the small-scale nature of production in the capital survived (and in many industries actually increased) throughout the nineteenth century, a fact which made the industrial (and therefore social) history of London so greatly different from other industrial and urban areas.[14]

The degree to which English economic and labouring life remained traditional can be gauged from the statistics of the census returns. In 1841, of the 6,700,000 employed only 2,619,000 worked in commerce, trade and manufacture. Another 1¼ million laboured in agriculture and a further million were employed in domestic service. In a number of regions, including London, going into domestic service (or taking work into the home) provided the obvious, and for some the sole, opportunity for female labour. It is true that, from the earliest days, the Lancashire and Yorkshire textile factories employed large numbers of female (and child) workers. But whatever *kind* of work women undertook it was always badly paid relative to men's. Female and child outworkers, however, were even more badly paid. The 'sweating trades', with their problems of low (and often decreasing) pay and excessive hours of work, became a major political concern later in the nineteenth century. Yet the basic outline of that problem and all its human consequences existed and was indeed striking throughout the years of this study.

It is generally accepted in the tortuous debate about the standards of living of working people in these years that, from the 1840s onwards, wage-earners embarked on a process of intermittent improvement in their purchasing power. It is also clear, however, that outworkers did *not* enjoy such improvements. When we recall that at the end of our period, 'outwork was still a regular feature of economic organisation and social life in several parts of the country, and the problems associated with it were accepted in a matter-of-fact way as normal, inevitable and unalterable',[15] it is obvious that we must qualify the impression of material improvements for working people by considering the adjacent circumstances of legions of outworkers. But the existence of substantial numbers of outworkers provides yet another illustration of the persistence of traditional features of economic organization. As with retailing and marketing, much of the labouring system of the new urban world would have been recognizable to earlier generations.

Historians concerned with the problems of the labouring people's standard of living in these years have paid a great deal of attention to their diet. For our purposes it is important to stress that the key changes brought about in plebeian diet stemmed not so much from economic fluctuations attendant on industrialization as from the changing nature of urban life. Urban working people developed a great reliance on specialized food producers and retailers, especially bakers and brewers. Any home-cooked foods had to be cheap and easily and quickly produced. Not surprisingly, potatoes, bacon and tea predominated. Moreover, all these foodstuffs could be purchased in small quantities, often on credit, at local shops.[16] As towns increased and grew in size they were able to serve the population through more and more varied shops, and it was possible to offer greater specializations of foodstuffs. Tripe and butchers' shops sold bacon, ham, pies, sausage and a host of regional and local foodstuffs – faggots, black puddings, brawn and haslet. Such shops were normally very local – only a very short distance away from the home – with the result that shopping tended to be local and frequent – often meal by meal. It was clearly an ideal system for low or irregular incomes. Yet, even in the provision of foodstuffs in the towns, we are again reminded of the persistence of old social patterns what continued to thrive in new urban habitats. Animals were driven into the centre of towns for sale and slaughter. Local butchers killed their own animals on their premises. Milk was taken from cows living inside the towns and cities, though from the 1840s onwards milk was also carried in volume from the countryside on the new railway networks.

London had long been dependent on a wide geographical area for its food supplies. And while most of the growing towns could feed themselves on their immediate agricultural hinterland, by mid century the big cities had begun to follow the example of London, Manchester, Birmingham, Leeds, Nottingham and perhaps a dozen other cities and were fed, to an ever-increasing degree, by the new railway system. Livestock for urban slaughter, and carcasses all arrived fresh on the country trains. Fresh fish was transported to any number of inland towns and cities. Milk first travelled to Manchester by train in 1844, and to London soon afterwards. Moreover, the trains lowered the cost of certain foodstuffs and thus increased consumption from the 1840s onwards.[17]

Despite the rapid growth of an urban population the nation was, on the whole, able to feed itself; only at the end of our period did it begin to develop its addiction to foreign foodstuffs – with all the consequences for economic life and strategic security. There were, it is true, hiccups in this process, none more troublesome than the hunger which proved so widespread (and fatal for many) in the crisis year of 1795. But even in that year hunger was nothing like as severe, nor was the associated unrest so volatile, as had been the case at regular intervals in the early eighteenth century. What is striking about these years is the nation's ability to feed itself. Given the rapid expansion of population, the concentration of that population in towns and the need of supply and retailing systems rapidly to adapt themselves to feeding labouring town's people, it is impressive how peaceful English urban life became. We know that the basic daily diet of the great majority of urban working people was meagre throughout these years. But that had always been the case. In fact, the ability to feed the growing population was testimony to the contemporary improvements in agricultural and farming life. And, as we have already seen, the close interdependency of rural and urban world is once more illustrated by the fact that many technical and agronomic improvements had initially been pioneered in towns. Often, however, although efficiency in agriculture brought benefits to the town dweller, the price was paid by rural labourers, large numbers of whom were driven from the land and forced to seek a living in the towns.

Women and children

By the mid nineteenth century the working population divided itself into three obvious categories. Some 30 per cent were adult males, one-third were women while children under 15 formed the remaining 35 per cent. About one-fifth of all the nation's women were in employment, but that figure ignores, of course, the economic role of women within the family and their labour in those activities privately organized around the family unit. This was particularly the case in agriculture and outwork. Of the 3.4 million women formally employed 1 million were in domestic service and another million in textiles and the clothing trades. And while it is true that there were no comparable numbers in any other industry, labouring women were to be found in almost every other form of industrial activity.[18] Of course female, like child, labour had been inescapable and undisputed features of labour since time immemorial, and it may therefore seem odd that they have both come to be regarded as one of the human scourges of nineteenth-century industrial change. Yet, to assert the obvious point of the antiquity of child and female labour is in no way to deny or minimize the appalling conditions endured by the early generations of industrial working people in a range of occupations, more especially in textiles, mining and pottery. The appalling evidence of these varied industrial inhumanities, marshalled and propagated by careful political propagandists as part of their attempts to ameliorate working conditions, were instrumental in shaping – and shocking – political and public opinion. And that in an age less shocked than our own by inhumanities. It is this self-same evidence which has left an indelible imprint on the collective national memory of the industrial revolution.

One unfortunate consequence of that impression has been to forget or minimize similarly appalling conditions both in the pre-industrial world and in the non-industrial sector.

Nowhere is this more strikingly the case than with domestic servants. With the general diffusion of prosperity, notably among the expansive middle classes, more and more girls and young women found employment as servants. The 1 million of 1851 had grown to 1.8 million in 1881 (and 2 million in 1914). In addition to forming the largest female occupational group throughout the nineteenth century, it was in many respects the harshest. While it is true to say that the women's labouring conditions varied enormously, it is clear, in general, that they were hired and employed on the most exploitative of terms. Yet they were, and remain, largely ignored or forgotten – though for obvious reasons.

> Immured in their basements and attic bedrooms, shut away from private gaze and public conscience, the domestic servants remained mute and forgotten until, in the end, only their growing scarcity aroused interest in 'the servant problem'.[19]

Ironically, however, they were often employed by people who were deeply concerned about the human and social problems of society at large but who often failed to notice the abuses conducted, quite literally, under their own roof; a striking illustration of that distinctive Victorian 'telescopic philanthropy' which could exhaust itself in distant causes but overlook outrages nearer to home.

The expansion of domestic service reflected the growth of the middle classes. Unlike past centuries when servants had populated the homes of the upper classes, in the nineteenth century they moved into middle-class homes, among 'the manufacturers and merchants, the bankers, brokers, lawyers, doctors and other professionals whose income depended, directly or indirectly, on industrialisation and the rapid growth of towns that accompanied it'.[20] Servants became more than the mere muscles of the middle classes' domesticity and ease. Over the century they were transformed into an important object of social status for those who could (or could not) afford to keep them. By the end of the nineteenth century the upper reaches of the working class came to be defined as those unable to employ a servant.

Servant girls had been traditionally drawn from the country, and many had learned the important rudiments of domestic service at an early age in the homes of local clergymen, tradesmen or farmers[21] before moving on to what they imagined would be the greater opportunities afforded by the towns, especially in London. This migration of female labour to the towns was accentuated by the spread – and cheapening – of the railways. The servant class expanded at something like twice the rate of the population as a whole. This growth was not merely a question of size, however, for there was a remarkable diversification of domestic skills as more middle-class homes became increasingly prosperous, and as middle-class social life became more complex. Later in the century, however, the rural supply of domestic labour began to dry up, while labour saving devices and the employment of 'daily' or hourly paid workers began to offer an alternative to the living-in servant.

In the first half of the nineteenth century the remarkable expansion of domestic service was a function not merely of changing economic patterns, but also reflected the new social styles gradually acquired by the self-confident bourgeoisie. The cult of Victorian domesticity demanded of wives and daughters managerial rather than labouring roles within the home. Furthermore, the architectural and furnishing styles so beloved of the Victorian middle classes were extremely labour-intensive. Large, tall houses, cavernous rooms, and an abundance of heavy furniture all required a great deal of time and attention in heating, cleaning, and general servicing. In the process a stark division of labour emerged both between the employing mistress and the domestics and, in larger establishments, among the servants themselves. Always carefully scrutinized by the household's womenfolk, subjected to a régime of unending toil (for the meagrest of returns) with little opportunity for rest and recreation – and even less for co-operation to defend their interests – domestic servants were to be numbered among the most oppressed of nineteenth-century working people. They were, at one and the same time, the instruments of Victorian propertied comfort and ease, and the victims of that same prosperity. If the new industrial proletariat made possible the material wealth of the employing and investing classes, their domestic counterparts guaranteed the creature-comforts and life-styles characteristic of Victorian bourgeois domesticity. And yet both groups had little to show for their efforts – certainly by the end of our period. Writing of the servant, Mrs Beeton noted, 'her life is a solitary one, and in some places, her work is never done'.[22]

Professions

The middle classes not only created employment opportunities for domestics but they were also the human material of the new and expanding professions. So familiar today, professional groups were essentially an offshoot of urban society in these years. It is true that towns had traditionally provided society with its rudimentary professional services, but the growing towns – and the middle and upper classes which came to control them (but did not own them) – required a variety of new professional services to provide for their private and corporate interests. The new urban and industrial systems not merely rendered many traditional professions inadequate – particularly as we have seen in the case of medicine – but they also created an urgent need for new professional skills, training and private societies to provide and regulate their practices. Medicine is, of course, the most obvious example of rapid and diversified professionalization in these years. In common with other areas, these changes came both from outside and within; from society at large demanding better treatment and control over medicine in all its forms, and from within as doctors, especially younger ones, sought to enhance their professional status and training, to exclude outsiders and purge medicine of some of its traditional practices.

The early nineteenth century saw the establishment of a series of new medical organizations and controlling agencies. In 1800, for instance, the Royal College of Surgeons was formed and in subsequent years a licensing system developed to regulate and control medical training, and teaching hospitals were established in London. In 1815 apothecaries were licensed and examined, and in 1823 *Lancet* was launched to provide a forum for serious medical and scientific research and the dissemination of medical ideas. As the century advanced there was a rapid proliferation of specialized medical literature and journals.

Parliament played a role in this reform of medicine. The 1832 Anatomy Act, for example, enabled students to pursue serious anatomical studies. But much of the reform of medicine took place at a local level, often under the aegis of local medical societies (some of older vintage). The BMA, founded in 1855, itself emerged from such provincial medical roots. It was, as we have seen, the devastation caused by the cholera in the 1830s and 1840s that illustrated the inadequacies both of contemporary medicine and the administration of social problems. Partly to cope with this the Medical Act of 1858 was passed, and despite its imperfections, it created for medicine a government organization and a code of discipline. By the end of our period medicine was clearly responding to widespread demands for more rigorous training and regulation. However incomplete and partial, these changes were instrumental in establishing medicine as a profession in the modern sense. It is also striking that a similar process was at work in other western countries.[23]

Much the same was happening in the ancient legal profession whose inadequacies, like medicine's (though perhaps less spectacular), were rapidly exposed by contemporary social and economic change. As with medicine, the important administrative reforms often took place in the localities, though culminating in the establishment of the Law Society in 1825.[24] Better-known still perhaps, because more obvious, and because they left an immediate and lasting monument to their work, was the professionalization of the clergy of all denominations. Not only did the English towns cry out for pastoral and spiritual care, but there was a vast and growing demand for missionary work abroad. And to satisfy that demand the surplus sons of the middle and upper classes – those with few prospects in business, trade, education or the armed forces – turned in growing numbers to the 'call'. The churches established their own colleges throughout England to provide a rigorous and professional training for potential ministers, and from these seats of divine learning armies of devout young men sallied forth to bring light and truth into the darker recesses of urban Britain, the old and new empires. Missionary work was available in abundance, whether it was in the East End, among West Indian slaves or their freed descendants, or in the perplexing challenges posed by India and, later, Africa. Whereas a century before the young John Wesley had ventured among the Indians of North America as a self-trained missionary, the early and mid nineteenth-century missionaries took to their unchristian peoples a rigorous and disciplined professional training which gave a sharper edge to their ambitions and goals. Their efforts were perhaps in greatest demand nearer to home, where traces of their activities are most obvious. In the first seventy years of the nineteenth century, 4210 new Anglican churches were built, along with some 20,000 dissenting chapels.[25]

The early years of this study were, however, little less than disastrous for the Anglican Church. Though its downward trend had long been in train, the demographic and urban transformation in England in these years rapidly accentuated its decline. Old parishes were by-passed by the drift of population; highly concentrated areas of population were ill-served (or not served at all) by the established church. By 1830 the Church of England 'was on the point of becoming a minority religious Establishment'.[26] This and related reasons led to the reform movement, beginning in the second quarter of the nineteenth century, one result of which was the rapid proliferation of churches and churchmen. From 1830 onwards there was a marked increase in men joining the ministry – and a sharp reduction in the problems of pluralism and non-residence that had crippled the church's earlier efforts. Moreover, the young men who went out to minister to the English people had, by mid century, brought to their role a new commitment which, in rejecting the old values and styles, was recognizably 'professional'. Gone was the 'hunting, shooting, fishing parson of the 18th century . . .'. In his stead was the religious professional; theologically sound, socially aware, and firmly committed to a rigorous enforcement of his religious and pastoral duties.[27] Like medicine and the law, the church had to reform itself at every level – of training, personnel, organization and even mentality – in order effectively to address its efforts to urban England. Without that professionalization, the church would quite simply have been incapable of conducting effective work among the English people. Thus, what may appear simply to be a matter of professionalization was (in the case of medicine, law and the church) a fundamental shift in direction and approach, as the agencies of middle-class skills (which were, of course, major employers of the middle class) sought to come to grips with the demographic and urban changes in English society.

No institutions had to make quite the adjustment required of the state itself. How to regulate, marshal and control the new urbanizing society and all its manifold problems was the central issue facing the inadequate machinery of the English state. Of course the demographic changes were only the most obvious of difficulties, and we also need to recall that the nation was involved in major wars in these years; against the Americans, 1776–83, and the French (with a short break) between 1793 and 1815. In common with the professional organizations, the governments of England required extra manpower to tackle these extraordinary problems. One result of this was a slow but steady increase in the civil service.

The 16,267 civil servants of 1797 (in wartime) had grown to 21,305 by 1832. In the next twenty years the ranks of the civil service almost doubled.[28] But it was only in the very last years of our period that reforms were first initiated which were to utterly transform the civil service in the following fifty years. Throughout these years much of the civil service was appointed by patronage with the sale of offices commonplace. It was, naturally, both inefficient and an object of contemporary criticism. Slowly, however, criticism led to reforms, first to curb corruption and, after 1828, to cut public expenditure. But the dramatic expansion of public business took place from the 1830s onwards, the bulk of it directed to the problems of urban life, as government created new agencies of control and administration, all of which demanded both the talent and the manpower to staff the emergent structure of administration. It may seem ironic, given the unreformed nature of Oxford and Cambridge (the natural suppliers of civil servants) and of the civil service itself, that the most striking and effective example of professional reforms was to be found in the administration of India. Administrators were trained at a professional college, they were subjected to competitive examinations and were selected by open competition. From the mid century onwards, these Indian reforms were to prove influential in the reforms of the domestic British civil service.[29]

Naturally the supply of manpower for these various middle-class professional occupations (and for other categories, for instance banking, business and insurance) came from the schools and universities. But these educational institutions were, if anything, in need of greater reform than any other contemporary organizations. Like the church (to which they were in many ways so closely related) English schools and universities had, by the late eighteenth century, fallen into decay and disrepute. The aspiring and economically aggressive middle classes required an education for their offspring that would train and equip them in a much more functional fashion than could possibly be achieved by the unreformed institutions of the late eighteenth century. Oxford and Cambridge, and the public schools that supplied their students, offered an education which was both unworthy of their traditions and unsuited to the changing demands of society at large. By the turn of the century, and along with other older professions, they all came under pressure to reform themselves. They began, slowly, to concede; reforming examinations, incorporating new physical sciences and encouraging a more professional approach to teaching and research. But their students continued to be drawn from the sons of the gentry and clergy and it was to provide an alternative that the University of London was founded in 1820s, for 'the youth of our middling rich people . . .'. London's curriculum offered a broad education, better suited to the needs of the times, though inevitably earning the scorn of those educationalists who saw the infusion of vocational and socially useful learning as a denial of the ancient traditions of learning. University College London was, in the famous words of Thomas Arnold, that 'Godless institution on Gower Street'. Within the new colleges of London University efforts were made to supply the professions with the necessary academic training. University College opened its medical school in 1828. Courses on the law were offered. Chairs of architecture were established at King's College in 1840 and at University College in 1841, while chairs of civil engineering were created at the colleges in 1838 and 1841 – a good thirty years before a similar development at Cambridge.[30]

The inadequacies and unattractiveness of contemporary education were best illustrated by the ancient public schools. Rote learning of classical texts, disorder and indiscipline, a sharp sense of social snobbery, all these and more rendered them unattractive to the middle classes. But the development of new public schools after the 1840s, made possible in large part by the development of the railway system, saw the early changes in public school education that were to enhance their appeal to the middle classes later in the century.[31] Already by the mid nineteenth century, however, the proliferation of new schools, and the reforms and revitalization of some of the old ones, created large numbers of openings for teachers. There thus evolved that distinctive and remarkable influential person of the modern professional schoolmaster.

It is possible to overstate the development of professionalism in these years. Despite the obvious and undeniable establishment of professional organizations, skills and training schemes, and the growing number of men working within them, significant sections of mid nineteenth-century English society continued to plod along, unaffected and unchanged, in a fashion similar to the early eighteenth century. This was strikingly so in the case of the armed forces which remained stubbornly immune to the benefits of reformed professionalism, reflecting instead the snobberies of an upper-class society which scorned and condemned the efforts and values of the middle classes.[32] After all, it was this military force which fought – and ultimately won – the protracted wars against revolutionary and Napoleonic France. But they were the armed forces of purchased commissions and of savage punishments of wayward soldiers and sailors. All this survived until the reforms of Lord Cardwell long after our period in 1868–74.

Throughout the second half of our period there emerged an unmistakable professional ideology in a wide range of middle-class occupations. In many respects it was an ideology that paralleled the changing discipline and work ethic among key sectors of the plebeian population. The ranks of the professional middle classes swelled enormously as demand for their services and skills increased with the expansion of the English middle classes, from whom the professional groups were overwhelmingly drawn and for whom, by and large, they worked. In the process they sought to secure their individual and collective interests by formalizing professional organizations, controlling membership, studies and qualifications, regulating fees and practices, and hedging themselves around with the structure of professionalism so familiar today. Central to their ideology was the commitment to a belief in merit – though a merit defined by ability rather than inherited social position or wealth. To detect and advance such merit the professions came to rely increasingly upon competitive examinations. Whatever their flaws, these examinations overcame the manifest inequities of the older traditions based on inheritance and birth. Moreover, the commitment to competitive examination was to become one of *the* central features of English society, in its government, administration and education, from that time forward to the present day. Indeed, this was to become one of the key ingredients in the modernization of society, for it drew a sharp line, often buttressed by the law, between the respectable and acceptable on the one hand and the unprofessional, the untrained – the quack or the imposter – on the other.[33] Increasingly, urban society came to be controlled and ordered by the professional classes and their corporate bodies. The labouring majority of the towns were, of course, regulated by different forces. The most potent of disciplines to which they were subjected was the workplace – however varied its form – where time-keeping, application and effort were not only basic to their toils but also of wider significance in the new urban habitat. Furthermore, many of the qualities essential in a working life were successfully and deliberately inculcated from within the family unit. After all it was in the plebeian family's interest that all its members should work for the family's material well-being. Harsh economic realities dictated to working people the obvious need to apply themselves industriously to their various tasks. The failure to do so – by accident or design – could be seen in abundance in the form of the urban poor. But whatever their fate, labouring people in urban England were overwhelmingly at the economic disposal of their immediate social and economic betters; the very people whose professional grip over urban life was tightening perceptibly at a time when working people had so little to show for their labours.

Earnings

The returns of a working life – the wages and salaries earned – were enjoyed unequally between the different social classes. Moreover, as industrialization advanced the distribution of income became even more unequally spread. The prosperous tended to increase their earnings; the less well-off, by and large, failed to do so. But this, as we might expect, is a complex matter which forms the heart of a continuing historical controversy which has raged for a whole generation; mainly about the standard of living of the working class in the first period of industrialization. For the more prosperous, non-plebeian sectors of the community there is little doubt that earnings and general material standards of living improved. Indeed, the life-styles of the expansive middle classes became a legend in their time, for growing material consumption and conspicuous displays of prosperity (to the disdain of their social superiors), and many of their social styles and architectural artefacts survive to this day. Throughout these years incomes rose for the middle and upper classes and some indication of this fact can be gauged from the payment of income tax – the number of people in comparable tax and income levels doubled between 1801 and 1848. In addition, their income increased by 52 per cent – in real terms an increase of 98 per cent. And a similar, if not more extreme feature can be detected the higher we move up the income scale. Such increases were in part a reflection of the shift in the distribution of incomes *away* from wages – itself a clear marker to the growing inequalities in earnings which was generated by industrialization.[34]

The most bitterly contentious theme in the argument about living standards concerns the impact of industrialization on the earnings and the living standards of working people. Not least, the contentiousness of this debate derives from difficulties about the evidence used. Assessing plebeian earnings is not as easily calibrated as the assessment of tax returns. Those historians who have opted for a bleak 'deteriorationist' view have sought their substantive evidence not merely in wage data but also in the more intangible aspects of social experience, often described as 'the quality of life'. Indeed, so complex and entrenched are the two visions of this phenomenon, 'pessimists' versus 'optimists', that it has recently been claimed 'it is probably fruitless to expect agreement about the evaluation of changes in living standards during the industrial revolution'.[35]

Whatever trends are discernible in real wages, it is clearly impossible to consider the standard of living – or quality of life – without establishing a critical formula that includes a range of variables. There are clearly enormous variations in working-class earnings across time, between trades and regions, seasonal shifts, and between 'booms' and 'slumps'. Then as now, wages were dramatically affected by the accidents and exigencies of personal and family life. Ill-health, old age, accident, marriages and large numbers of children; these and many other variables could greatly alter the material well-being of the individual or family. Moreover, when we recall that old age so often plunged working men and women into poverty, thus offsetting whatever material benefits a labouring life had produced, it becomes clear how difficult is the task of reconstruction.

Of all labouring groups examined here, there is little doubt that, throughout the bulk of our period, low wages remained central to the various social problems. During the 1790s – the outbreak of war with France – there was a significant fall in real wages, though they are thought to have risen again, immediately after the war (1815), and again in the 1840s. But this was of little consolation for those growing legions of people who were either under- or unemployed.[36] Women's wages (and children's of course) were markedly low, and much lower than men's throughout these years. And since child and female labour comprised perhaps three-quarters of factory labour and a substantial majority of *all* wage-earners it is scarcely surprising that such work was characterized by low pay, despite improvements after 1815. Similarly, there seem to have been improvements in the earnings of other groups closely and directly involved in industrial change: miners, transport workers, shipbuilders and others. We need to remind ourselves again, however, that such factory and 'heavy industry' workers were greatly outnumbered by other labouring groups. And it was 'the others' who signally failed to benefit (and in many cases were actually harmed) by the process of industrialization. Nowhere was this more strikingly so than with the handloom weavers who declined from their buoyant years in the 1770s to 1790s to a pathetic state of marginality by the 1830s. Within two generations they were driven from a proud and prosperous trade to the level of abject poverty. Bolton weavers earning 25/- a week in 1800, scraped only 5 to 6/- by 1829.[37] Their standard of living 'at best stood still, and at worst deteriorated sharply'.[38] The only redeeming feature of their story is that they greatly declined in numbers.

The number of domestic servants, as we have seen, greatly increased. Yet their earnings too were a by-word for low pay. Naturally, domestic pay was highest among the elevated of that highly-structured society: house stewards, chief chef, butler, valet, footmen, coachmen and housekeeper. But even at this rarified level of domestic attainment the wages were low; a housekeeper might earn 24 guineas a year and a butler 50 guineas. At the humbler level, the meanest of domestics might be paid only 8 guineas a year. In addition they generally received clothing, food and accommodation, but to be set against their income was the inescapable scrutiny and ceaseless chivvying of employers and senior domestics. Compared with factory or outworkers, domestics were thought to be well provided for; by 1867 it was calculated that 258,000 had bank accounts with an average of £28 each. For many, however, their earnings were inadequate as a decent family income, and in any case they were normally obliged to live away from their families and children.[39]

Any such discussion about earnings and standards becomes unreal when we divert our gaze towards the very bottom of the social order; to those legions of people who, with no regular income or job, scraped whatever precarious living they could from the unyielding surface of urban life. Most notable perhaps in London, there were knots of peripatetic salesmen, beggars, crossing-sweepers, mudlarks, urchins, whores, entertainers, errand boys and girls, labourers, and a multitude more, all jostling for survival in a hostile environment. It was the wretchedness of their lives reported by a growing band of social investigators which, by the end of our period had come to attract a great deal of social concern and political attention. They were, in human form, the problem of poverty itself; an aggregate human suffering which remained a growing political preoccupation throughout the century. And however persuasive the historical arguments might appear about the gradual upturn in material prosperity wrought by industrial and urban change, however convincing the case for an enhancement of real wages for ever more working people, we will never grasp the totality of urban experience unless we confront the existence and depth of contemporary urban poverty. Not unlike our own times, the undeniable fact of improvements for some did not (and cannot) erase the contrary historical experience; of generations of people locked into an inescapable spiral of personal and family destitution which seemed immune to whatever improvements are clearly detectable in other sectors of society.

4 *The language of class*

The powerful demographic and economic forces responsible for the physical and human transformations in England in these years were paralleled by changes in the way in which the English viewed themselves, individually and collectively. Indeed it became a source of major intellectual endeavour in the course of the nineteenth century to plot the changes in these perceptions; a pursuit that haunts historians of the period to this day. Yet the search for an accurate description or analysis of English society did not begin in these years. From the late seventeenth century onwards, beginning with Gregory King, there were periodic studies of English social structure.

Throughout much of the eighteenth century, and up to the beginning of our study, social analysis tended not to speak of class but rather the 'upper orders', 'the middling ranks' and 'labouring men'. Furthermore, there was a tendency to define groups more in terms of interests, or 'wealth, occupation, region, religion, family, political loyalty, and connexion'. In the words of the most recent historian of the eighteenth century, 'Struggles were less between "class" and "class" than between "ins" and "outs".'[1] By the 1840s, however, a change had clearly begun to make itself felt and the language of social class had begun to establish its distinctive claims.

By the 1830s and 1840s radical politicians frequently spoke of social class as *the* central issue in contemporary political debate and suggested that contemporaries, increasingly, viewed themselves within the collective identity of their own social class. Even before Marx established the intellectual analysis of class, contemporary activists returned, time and again, to the same question. Social conflict – as opposed to the analysis of class – had of course long been in existence. The popular culture of eighteenth-century England often enshrined a collective organization, *esprit* and rebelliousness which gave both common expression to plebeian sensibilities, and taxed and threatened local men and agencies of property and control. Often such clashes were distinctively local. None the less, there is a detectable pattern here; of plebeian action, in a myriad of cultural forms, which probed the limits and weaknesses of the governing system administered by the gentry.

The political dominance of the gentry was not, however, seriously threatened by these varied (and, to repeat, local) plebeian responses, for they were overwhelmingly constitutional. In the words of E. P. Thompson, they were 'aimed often at recalling the gentry to their paternalist duties'.[2]

This pattern was broken in the decade of the 1790s by the impact of revolutionary, notably Paineite, ideas which transformed not only the ideology of popular resistance but also provided a new, popular framework and organization that sought to challenge and transcend the existing political order. One of the key and innovatory features of popular radicalism of the 1790s was its urban nature. Whereas the strength of English radicalism in, say, the 1770s and 1780s had been among the propertied gentlemen of the counties (in conjunction with older metropolitan forces), the popular radicalism of the 1790s was an overwhelmingly urban phenomenon. Its centre of gravity was among artisans and tradesmen in London and it had significant support in a number of other provincial towns; in Sheffield, Manchester, Norwich and in Scotland. Above all else it was both urban and plebeian, factors which distinguished it from earlier radicalism (and made it more troublesome). How to control it – more especially how to limit its growth in provincial towns where central government lacked effective organization or agencies – proved to be one of the government's most worrying concerns in that decade. The unprecedented strength, ubiquity and vocabulary of the popular movement of the 1790s was only kept under control by a régime of encircling and prohibitive intimidation and repressive laws, many of which were to prove useful, once again, after 1815. But it was from the ideals and organization of the 1790s that the revived working-class agitation of the post 1815 years was to flower. It is at this point – when the urban ideology of the late eighteenth century breaks surface into the transforming world of post-war upheaval, infusing itself into the general urban (and in some measure industrial) unrest – that class becomes more obvious, more instantly recognizable. In the first instance, however, it was not the emergent working class which registered the most effective presence in these years, for it was in these same post-war years that the rising middle classes themselves became increasingly self-conscious and politically self-aware. In response to the ever more strident middle-class political

cal position and voice, more and more working people turned to their own political and trade organizations to safeguard their economic interests and advance their precarious political claims. In its turn, the radical popular movement – its ideals and vocabulary largely created in the 1790s – is what many historians have turned to for an explanation of the emergence of working-class consciousness, most notably in and around the Chartist agitation of the 1830s and 1840s.

To understand the nature of the emergence of nineteenth-century class society we are, once more, forced to come to terms with the society from which it emerged. It is tempting to consider that early and mid eighteenth-century England was governed by a tight-knit caucus of aristocrats – that 'federation of country houses', with extensive landholdings, homes in the capital and power and influence wielded both in the Lords and the Commons. Corruption was the lubricant of the system; bribery, inducements and preferments offered for services rendered or promised and best administered and remembered in the person of the Duke of Newcastle. Yet political influence and landownership were not the unique preserve of the old aristocracy. Land, often owned in substantial holdings, had been acquired by the middle classes, whose wealth had grown from the seventeenth century onwards through a complexity of trade, professions and businesses. Tradesmen, merchants, lawyers, the very great majority of them town dwellers, and the beneficiaries of Britain's expansive overseas empire – the shippers, brokers, suppliers and mercantile houses, planters, colonial officials, military personnel, and the outright colonial speculators all (and many more) sought to secure their economic base and their social standing by purchasing land. This was perfectly understandable because, whatever its social structure, the pinnacle of English society was undoubtedly dominated by an aristocracy which owned and controlled vast tracts of the nation's land. Furthermore, their dominance of land increased during the eighteenth century. Whereas in 1700 they owned between 15 and 20 per cent of landed wealth, a century later this had grown to 20–25 per cent. Their wealth was to be enhanced even further by the simple possession of land because they dominated substantial parcels upon which the towns and industries of the late eighteenth century were developed. They were the 'sitting beneficiaries of industrialisation'.[3] And yet, notwithstanding their wealth and status, by the mid eighteenth century the traditional aristocracy had begun to envy the prodigious and flashy materialism of the *arrivistes*, notably of returning nabobs and planters.

The wealth and prosperity generated by British imperial and maritime trade had become legendary. Indeed, we need only look at the physical transformation of the major international ports – London, Bristol, Hull, Liverpool and Glasgow – to see surviving material monuments to the prosperity generated by eighteenth-century long-distance and imperial trade. Often too this prosperity was channelled into landed estates. In the 'season' such men gathered around the fashionable pleasures of Bath, as Smollett's Squire Bramble bemoaned.

Every upstart of fortune, harnessed to the trappings of the mode, presents himself at Bath. . . . Clerks and Factors from the East Indies, loaded with the spoil of plundered provinces; planters, negro-drivers and hucksters, from the American plantations, enriched they know not how; agents, commissaries, and contractors, who have fattened on two successive wars, on the blood of the nation; usurers, brokers and jobbers of every kind; men of low birth and no breeding, have found themselves translated into a state of affluence, unknown to former ages . . . and all of them hurry to Bath, because here, without any further qualification, they can mingle with the princes and nobles of the land.[4]

It was among such men that the advent of modern capitalism was to be found. By the late eighteenth century there were unmistakable signs of a capitalist class – in rural and urban society – using the efforts of a propertyless workforce, itself able to offer nothing but its labour. This was a process which was most striking in the late eighteenth-century textile industries – *before* the coming of the factories – where substantial capitalists effectively employed small armies of domestic workers on an outwork basis.

While Bath proffered its more beguiling if transient pleasures, for the *nouveaux riches* the purchase of land provided them with something more substantial and durable; political influence and power. It is, for example, instructive to recall that one of the newer political groupings in eighteenth-century Parliaments were the planters from the West Indies and (until 1776) from North America. This 'West India lobby' formed a wealthy and powerful group, both inside and outside Parliament, throughout the eighteenth century. Significantly, however, their power rapidly evaporated after the reform of Parliament in 1832, followed six years later by the ending of their economic power with the emancipation of the slaves.[5] And yet they were relative newcomers; *arrivistes* able to secure a political toehold via land, thanks to their successful economic rapaciousness in the Atlantic economy. It is true that they and many others were quite unlike the subsequent Victorian middle class and *nouveaux riches*. For our purposes, what is important is the planters' consolidation of political influence through land in the mother country. Land – and its associated political power – was quite clearly not the narrow monopoly of a traditional aristocracy.

Aristocractic sway and political influence seems to be more overweening when viewed from London. Moreover, political power often looks quite different when we examine *local* politics. Indeed, it is worth reiterating the crucial fact that England was governed primarily, not from London but in the localities, through that varied, confused and, more often than not, archaic and ramshackled structure of local government inherited and added to since time immemorial. Manchester, for example, in the 1790s, despite its rapid physical expansion and political volatility, continued to be governed by the remnants of the medieval Court Leet. Throughout the country it was the system of the vestry, its origins lost in time, that provided the backbone of local government.[6] Here, and on the magistrates' bench, local men of substance administered the agencies of control and relief: the poor law, the militia, raised local taxes and sought to dispense with local nuisances. From the mid eighteenth century, as urban society's problems multiplied, there proliferated a host of local improvement commissions, all trying to cope with the swelling tide of physical urban problems.[7] And, at most of these levels, local administration was not directly in the hands of an old aristocracy.

This becomes clearer when we examine the magistrates. By 1831 almost one-quarter of England's magistrates were clergymen, not surprising perhaps since they were often the only men with education and property appropriate to the task. They appeared, none the less, to be in the pay and the pockets of their betters; the local stooges for invisible, though much more powerful social groups. 'Too often they seemed to be administering rich men's justice . . .'.[8] Naturally enough, local government was itself changing – a further symptom of the shifting contours of economic and political power. But whatever particular form local administration developed, behind it lay the weight of property, sometimes (notably in the countryside) in the form of extensive landholdings. In towns, political power tended to be in the hands of 'substantial traders and professional men, often with a stiffening of local landowners'. There was, then, a marked difference between the kind of men who governed in the urban localities and those who dominated the offices of central government, for there can be no doubt that the presence of the aristocracy overshadowed all the major offices and institutions of the state. As Harold Perkin has shown, the landed class dominated the Commons until 1885, the Cabinet until 1893, it controlled entrance into the Civil Service until 1870, the army until 1871, and the church for much longer. And it widely and persistently overshadowed local government throughout the nineteenth century.[9]

Despite the shadow of landed aristocracy, effective local control passed ever more clearly into the hands of the emergent middle class. As progressively more businessmen made their local mark, they were gradually incorporated into the offices of local administration, and in their turn they came to impose on local government and its agencies a style and mentality in keeping with their business background. By the late eighteenth century there was a discernable amalgam of old and new political and social interests in the government of towns, as the old landed and new business groups sought to govern effectively in their own collective interests. Later still, especially after 1835, the early nineteenth-century business interests were able to come into their own. Indeed, the physical and architectural monuments, which dot the Victorian urban skyline to this day, bear testimony to the ideology of the Victorian business élite transplanting its ideals, ambitions, and style, into the administration of local government and the construction of the Victorian city. This was a transformation greatly aided by the legislative reform of local government, notably the 1835 Municipal Corporation Act which opened the doors of local government to men of substance, especially the nonconformists who had for so long been kept outside English politics. In the years after 1835 many of the towns fell into the political lap of the middle classes.

This was not only a fair reflection of the shifting economic power within urban life, but also a short-term culmination of the increasingly strident middle-class demands for a greater political voice. Indeed, it was among these very groups that historians have seen the first effective expressions of the modern language of class. If the 1835 Act was a success for the middle class, it was a success born of political efforts going back to the eighteenth century.

Once more, the crucial events took place in the 1790s. In those years of political turmoil there was a slow coalescing of the propertied orders against what they took to be the subversive threat of Jacobinism. The rise of popular radicalism between 1791 and 1795, articulating Paineite demands for a broadly-based democracy, in a great variety of regional and plebeian accents, and all under the shadow of revolutionary France, transmuted radicalism into Jacobinism, at least in the eyes of its propertied opponents. Property itself seemed threatened by the advance of the rights of man. Edmund Burke became the scribe of the counter-revolution; the rallying point for those, inside and outside Parliament, who viewed popular radicalism with alarm. It is symptomatic that the political arm of anti-radicalism took the title 'Association for the Preservation of Liberty and Property against Republicans and Levellers'.[10] There were exceptions, of course, but by the mid 1790s the propertied orders – aristocrats and middling orders – had closed ranks against their plebeian and radical opponents. Burke and a galaxy of lesser-known writers (often evangelicals) began to promote the concept of an alliance, a coalition of all propertied classes, to face that hostile federation of the propertyless and voteless pressing menacingly on the outer fences of the propertied bailiwick.

The newer middling orders had begun to express an economic voice as early as the 1780s when feeling themselves threatened by government fiscal policies. Throughout the last twenty years of the eighteenth century there had been regular pronouncements about the invaluable economic influence of the middle classes, and it was, on the surface, hard to deny their own claims of having brought ever more material prosperity to the nation. In the 1790s this middle-class voice was strengthened and inflated by the general regrouping of propertied interests faced by what they considered to be local Jacobinism. It was, however, a political and ideological unity which was itself transmuted into a new form of class identity by the war fought against revolutionary and then Napoleonic France. The war also gave the middle classes and their literary supporters grounds for complaint. As the fighting dragged on, they alone felt themselves to be financing the struggles; complaining of 'the burden of taxation [which] presses so heavily on the middle classes of society . . .'.[11] In fact, many of the indirect wartime taxes fell disproportionately on the poor. But Pitt made efforts to tax those more easily able to bear the costs, offering a clear ideological as well as a financial justification: 'in a war for the protection of property it was just and equitable that property should bear the burthen'. In introducing his house tax, Pitt devised a scheme which 'might be supposed to bear a more certain proportion to the fortunes of the individuals'.[12]

The century drew to a close (as it had begun) with the nation locked into a fierce and all-consuming war. Symbolically perhaps for the future, in 1798–9 propertied society found itself faced by Pitt's new proposal – income tax, applicable to incomes over £50. But at 2/- in the pound, it smacked, to its outraged opponents, not only of yet another gross attack on property, but a pale imitation of French principles, against which the wars were, in theory, being waged.

> This measure [said George Tierney] puts a tenth of the property of England in a state of requisition, a measure which the French have followed, in their career of revolutionary rapine. . . .[13]

At the turn of the century, the propertied classes (especially in the lower reaches) were undoubtedly feeling the pinch, their bleating complaints recurring time and again in contemporary accounts. One man supported income tax because 'the remainder of his property was secured from the inroads of French pikes and French principles'.[14]

Peace in 1815, and the abolition of income tax and property tax in 1816, brought only marginal relief, for the financial problems of the war years were rapidly overtaken by, in many respects, the greater difficulties of post-war recession. It was in these years that the language of class began to take on a more specific and divided form. The most striking feature of post 1815 politics were the fractures within the wartime coalition of propertied interests. Changing economic circumstances compounded by political change (notably by the 1815 Corn Laws to safeguard the agricultural interests) served increasingly to divide the new manufacturing groups from the landed interests, while simultaneously convincing revived plebeian radicals of their social and political isolation. The fissures of a new society – divided into classes – were becoming wider and more visible.

Although the political turmoil after 1815 marked a sharp veering away from the wartime consensus among propertied society, the social divisions which emerged after 1815 were unmistakably shaped by wartime experience. The middle classes had been transformed by the war; British industry had flourished and buoyed along by the increasingly powerful commitment to free trade throughout the world had greatly outstripped any potential competitors. Working people fared much less well, however, and the 200,000 demobilized soldiers and sailors drifted home to contracting industries. Not surprisingly, many ex-servicemen turned to local radical politics, propelled in that direction by the bleakness of economic circumstances and influenced by local traditions (which many of them had missed) of political and industrial agitation. There was in addition the persuasive radical demand that men who fought for their country ought to have a say in its government.[15] Unfortunately, they faced government ministers who had served their political apprenticeship under William Pitt in resisting French principles. Men who had bitterly resisted political change in the 1790s proved no less obdurate in the years after the war.

It was against this impenitent official refusal to consider political change that plebeian and middle-class reforming movements were to swirl for the next three decades. In the process, the political movements spawned by the different social groups were to feed upon and compound the growing awareness of social class. They were to become ever more aware of social separation and distinctiveness, one against another. Social class – and talk about social class – soon established itself as *the* unique and overweening characteristic of English society. It is, then, in these post-war years that we begin to detect the open development of a class consciousness which, in time, came to shape the political experience of growing numbers of working- and middle-class people. This is not to deny the remarkable local, trade, and temporal variations in such experiences – of the marked gradations and nuances within each social class. But the existence of such variety does little to undermine the centrality of social class, both as a methodological tool enabling us to make sense of nineteenth-century English history and, more importantly perhaps, as near an approximation as we can detect to social experience as it was lived.[16]

It is widely agreed that the radical and plebeian politics of the years 1815–48 played a crucial and determining role in sharpening the lines – and enhancing the language – of social divisions. There are, inevitably, arguments about which years are more especially important; the post-war years between Waterloo and Peterloo, the struggle for 'The Bill' up to 1832, the anger and despair following the Poor Law Amendment Act of 1834, or the years of Chartist agitation in the 1830s and 1840s. In all these phases it is relatively easy to show the contribution made by radicalism towards the formation of class, but this is far from claiming that all working men were radicals. Conversely, not all radicals were working men. As we have already seen, the ideology of urban radicalism after 1815 had been shaped during a much earlier period and did not, in any sense, belong to one particular social class. It has been shown by recent historians that the ideology of radicalism could appeal just as easily to the provincial shopkeeper or country squire as to the urban working man.[17]

Radical ideals clearly appealed to large numbers of working men and to many of their social betters. The fact that they were unable to unite on these principles is to be explained by the widening gaps of social class and the swift realization that their self-interests were, in many key respects, mutually contradictory. And this central fact was illustrated – and shaped – by the two major political movements between 1815 and 1848 that came to provide polar opposites in national politics; the middle-class campaign against the Corn Laws and the working-class Chartist movement of the 1830s and 1840s.

The campaign against the Corn Laws of 1815 (which were so blatantly in the interest of landed society and against the economic and political interests of working- and middle-class people) formed the basis of a new middle-class political agitation. The campaign against the Corn Laws was a long haul, advanced by the Anti-Corn Law League (founded in 1839), and culminating in the repeal of the Corn Laws in 1846. From the first to last, the League was a middle-class body, advancing what it took to be middle-class economic interests by persuading groups in Parliament. True, it did have some working-class supporters and, ultimately, wore down the aristocratic and landed opposition. In the process, however, the League was instrumental in shaping middle-class political consciousness. Cobden later confessed; 'We were a middle-class set of agitators [using] those means by which the middle class usually carries on its movements.' When repeal was passed, in 1846, its opponents spoke openly of the success of the middle class. Cobden asked Peel; did he 'shrink from the post of governing through the *bona fide* representatives of the middle class'?[18] In conception, style, composition and ambition, the Anti-Corn Law League was an instrument of the middle class. Again, to quote Cobden; 'I am afraid that most of us entered upon the struggle with the belief that we had some distinct class interest in the question.'[19] The League was avowedly propertied, exclusive and self-interested, and not above distortion to advance its sectarian case.

The Anti-Corn Law League spoke specifically for the middle classes. For that very reason it was an object of distrust to the movement which, in its turn, was instrumental in shaping working-class political identity – Chartism. Chartists had mixed attitudes towards the League (largely because of their ambivalence towards the Corn Laws themselves). At the local level, however, as working- and middle-class activists battled for local political dominance, there is little to be seen but friction, social and personal antagonism, and a legacy of bitter words. It was a story of middle-class Leaguers opposed by working-class Chartists. In Sunderland for instance, Chartists asked Leaguers,

> What is our present relation to you as a section of the middle class? . . . It is one of violent opposition. You are the holders of power, participation in which you refuse us. . . .[20]

A similar tale unfolded across the country. London Chartists claimed, 'our cause, which is the people's cause, is not the cause of party or faction, but the cause of the great mass of industrious classes. . .'. They denounced the seductions of the Leaguers; 'a party comprised of avaricious, grasping, money-mongers, great capitalists and rich manufacturers'.[21]

Wherever we look, the fundamental clash between these two organizations took the form of class antagonism. There were efforts in 1841–2 through the Complete Suffrage Union to bring both sides together in a joint movement, to effect a 'reconciliation of the middle and labouring classes'.[22] But it was doomed to break up on the ever-sharpening personal and collective animosities of one side towards the other. Chartism, like the League, was openly rooted in, attractive to and advancing the interests of one particular social class. A recent historian of Chartism, surveying the full range of its surviving evidence concludes,

> The tone of a range of historical sources, from newspaper accounts (friendly and hostile), Home Office reports and criminal records to NCS Executive lists, the autobiographies of contemporaries, is working class. . . .[23]

It is true that working-class support for Chartism was uneven and fluctuated during the 1830s and 1840s as trade and employment rose and fell. Indeed, it was from the social turmoil – and political resistance – created by the down-turns of the economy that Chartism drew its strengths. It tapped plebeian misery and offered the distant prospects of improvement through political change. In the process it drew upon and then, in turn, reinforced the emergent sense of class isolation. Chartists and their armies of backers saw, and portrayed themselves confronted by, an aristocratic and landed Parliament, resistant to their legitimate political claims, and a middle-class of industrialists and capitalists whose own political lobby, the League, seemed to threaten working-class well-being from a different perspective. Thus, the crisis years of the 1830s and 1840s witnessed the apogee of class politics in our period.

Working-class politics was particularly instrumental in forging the language and the awareness of class. Throughout, there ran the central demand for parliamentary reform, a demand which reached a crescendo in 1830 when it coincided with revolution in France, widespread unrest in rural England and mounting industrial disturbances. The years 1830–2, with their range of reforming campaigns across the country, formed a major political crisis. Demands for factory reforms, trade unionism and above all the demand for 'The Bill' created a turmoil of organized radical politics. Throughout that agitation the language of class was paramount. Although the campaigns for parliamentary reform in those years were characterized by co-operation between middle- and working-class radicals, the act of 1832 splintered their tactical unity. Working-class radicals reeled away from the act, which gave them nothing, infuriated and betrayed; let down by both Parliament and by their erstwhile middle-class colleagues. But the working-class reformers wanted more than the vote; they also wanted a free press. Instead, popular publishers and journalists found themselves harried as never before, as the government sought to staunch the flow of radical ideas to working people from the presses. To compound this working-class agitation, the early 1830s also witnessed a major revival of trade unionism.

Perhaps the most pressing of working-class political issues was that concerned with working conditions in the factories – the 'Ten Hour movement'. Although the factory system was still small-scale and geographically restricted, the problems it harboured were appalling – and the stuff of plebeian discontent and agitation. Throughout, the bitter complaints of working people, even when expressed by middle-class supporters, accentuated still further the class divisions in political life. This tendency reached new heights of bitterness and hostility in the opposition to the 1834 Poor Law Amendment Act.

Working-class politicians needed no prompting to point out that the 'great betrayal', the act of 1832, had been swiftly followed by the most overt of class legislation. The organization of the Ten Hour movement was rapidly transformed into local opposition of the Poor Law. And there followed a bitter fight against its local implementation which was sometimes successful in gaining modifications. But what accentuated working-class political frustrations was the parallel and contrasting success of middle-class politics. Three Acts of Parliament in particular seemed to confer extraordinary local power on the middle classes. The 1828 Repeal of the Test and Corporation Acts removed the final political disabilities from the Dissenters, who had often acquired local economic and social prominence. The 1832 Reform Act was thought to enfranchise the middle class – although in fact it was nothing like as extensive as was claimed. But most important of all, the 1835 Municipal Corporations Act broke the political monopoly of old urban oligarchies, allowing new men – of the middle class – to secure a political grasp on the cities. By the late 1830s, the middle class had secured local political dominance – as councillors, aldermen, magistrates and Poor Law Guardians. 'The economic and social masters of the working class had emerged also as their political masters',[24] although there were, of course, great local variations in this political story.

By the mid 1830s, then, there had come into being a heightened awareness of working-class identity, born, in the long term, of the political agitations stretching back to 1815 and nurtured, in the short term, by the bitter experience of 1832 and after. Working men and their political leaders looked in frustration at the contrary fortunes of their local betters, who were now not only economically dominant, but also in full control of a wide – and expanding – range of local government and administration. What, on the other hand, had working people got to show for their efforts? To the informed working-class communities, all the effort and troubles of twenty years had yielded naught for their comfort. By the mid 1830s there were precious few signs of material benefits for working people. Town life was, in material (in domestic and medical) terms, bleaker than ever and there was no amelioration (for those involved) in industrial working conditions. A working life contained its own unavoidable pitfalls and the outcome was all too often a decline into the poverty of old age. And over all this, there hovered the spectre of the workhouse. Not surprisingly, plebeian communities throughout urban England exhibited signs of frustration and anger. It was from this heady political context that Chartism was to emerge.

Based on traditional radical principles, with a national network of supporters and activists, and able to muster millions of names to its petitions, Chartism is widely accepted as the first working-class movement in modern British history. It was also much more besides. It was Chartism's working-class composition, style – and threat – which impressed its supporters and alarmed its opponents. It is clear beyond doubt, as its enemies in any case knew and feared, that 'the Chartists were representative of the working populations of their home communities – in fact they were "ordinary" working men'.[25] There were, admittedly, spasms of violence; strands of turbulence woven into the complex fabric of the Chartist movement. But the sharpness of its language, the depressed anguish of its followers (notably at those time of trade recession which drove them towards Chartism) cannot hide the fundamental peaceful and constitutional nature of the movement. Chartism's greatest threat came not from its violence but from its fundamental democracy. It is also true that other radical and working-class movements competed with Chartism for the hearts and minds of working people, most notably perhaps the Owenites and the proliferating trade unions. But it was, and is, the Chartists who epitomized and advanced working-class politics. And in sharp social and political opposition there stood the political arguments developed to advance the interests of their middle-class opponents – the Anti-Corn Law League. Both expressed the interests of their constituents and both justified their politics in terms of social class.

What made working-class politics so bitter – so sharpening the edge of their language and style – was not only the economic plight of the 1830s and 1840s, but the contrasting fortunes of their political opponents. After the rejection of the third Chartist petition in 1848 (by a government whose resistance was stiffened by the sight of European revolutions), Chartism's working-class supporters had gathered few political crumbs from the tables of their betters, despite all their efforts over a generation and more. Their middle-class opponents on the other hand ruled the roost. They appeared to have gained control of the local administration of towns and cities. It seemed that they had won a series of major legislative concessions. They had, or so it seemed, secured the Repeal of the Corn Laws. And now, in conjunction with older political vested interests in London, they had effectively beaten back working-class demands for political reform.

The paradox remains however that the bitterness of politics in the 1830s and 1840s was very soon dispelled or submerged. By the end of our period it seemed that a political calm had settled on the land. This was largely because of the economic upturn, and yet, notwithstanding these new-found economic fortunes, the language of class remained. By mid century the existence of social class and its corresponding reflection in daily and political vocabulary was everywhere accepted. In 1850 Palmerston put the case succinctly:

> We have shown the example of a nation, in which every class in society accepts with cheerfulness the lot which Providence has assigned to it.[26]

True, Palmerston also thought that this 'cheerfulness' was shaped in part by the prospects of social mobility. We now know that it was a prospect which became progressively more distant the lower down the social scale we look. None the less, Palmerston's words are instructive for, by mid century, there was no doubting the presence – indeed the pre-eminence – of social class.

What had made the post 1815 radical and industrial movements so potent and so worrying a force was the habitat in which they thrived. In the 1790s the centre of gravity of radical politics had remained with the London Corresponding Society in the capital. After 1815, however, that centre of radical gravity had shifted – to the expanding urban and industrial areas (not always one and the same thing). Between 1815 and 1851 it was *provincial* urban life which shaped the direction and style of radical politics, notwithstanding the need of those movements ultimately to direct their complaints or demands to the fount of ultimate political power in Westminster. It was this provincialism of early nineteenth-century popular politics that made the national governments' problems so acute. They were, increasingly, governing at a distance, through their friends, associates or agents many hundreds of miles away from London.

To make matters worse for national government it was the unique nature of provincial urban life which made political life so uncertain and which in turn was instrumental in shaping the nature of English social class. Although early nineteenth-century cities and towns varied greatly, they were often characterized by sharp social segregations and, in many cases, were overwhelmingly proletarian. Historians and social geographers continue to argue about the degree to which urban areas were segregated; about how far their different and distinct divisions and suburbs provided a home primarily for one social group to the exclusion of others. To put it crudely, middle-class families tended not to live in working-class communities and vice versa. More recently it has been argued that Victorian towns were in fact much more socially varied – that suburbs contained an unusual social mix – but it is likely that these arguments might apply to specific towns (especially Leeds and Birmingham).[27] What is striking is the fact that contemporaries – and the great majority of recent historians – have argued quite the opposite; that the Victorian town was segregated. For our purposes it seems likely that the segregation of the English town was an important ingredient (though varying in its local impact and direction) in the evolution of social class. It was the physical nature of local urban life which played so important a role in shaping both the nature of social life and in providing the framework within which collective self-awareness emerged.

In many northern and Midland towns, where working-class communities developed in relative social isolation from their social superiors, there developed the political and trade organizations that served to compound the sense of separation and collective difference. With the proliferation of formal political movements, working-class organizations were often pollinated in other plebeian communities. There was in a sense, and especially in the 1830s and 1840s, a contagion of radical plebeian agitation that spread throughout urban England – even to towns relatively untouched by social segregation. The development of class awareness was not after all caused by any single factor. It seems to have been a product of a complexity of forces, often different from one area to another. Thus, the *physical* shape of the local urban area – its degree or lack of segregation – will not always provide us with the crucial determinants of the development of class-consciousness.

A crucial factor was the impact of radical or trade organizations. Indeed, it was part of their self-avowed purpose to *generate* a national commitment and organization among their plebeian followers, though this was often inordinantly difficult and they often stumbled over the peculiarities and details of *local* interests and ambitions. By their very nature, radical and plebeian movements – from the corresponding societies through to the Chartists – were missionaries, spreading the views among their peers, by the well-tried and often traditional methods of propaganda and agitation, in order to create interest and collective commitment. In the process there evolved a federation of plebeian political communities which, at certain crucial junctures, were transmuted into a *national* political movement – but with their organizations firmly rooted in local experience. It was these plebeian movements that were the forcing house for the development of a class-consciousness which transcended the particularities of local plebeian trade or urban life.

Having said all this, the modern historian cannot help being struck by the remarkable differences between Victorians towns and cities; differences of geography, economic structure, shape, demography and local political experience. The process of urban and industrial change served not to make English cities the same, but to make them different, as Asa Briggs remarked; 'to differentiate English communities rather than to standardize them'.[28] Urban communities were greatly different and so too were the experiences of their inhabitants. By examining the major political movements (themselves responding to these varied urban peculiarities) it is seductive to detect among their followers a more clearly defined collective and national sense than might have existed at the more mundane level of daily experience. None the less, it would be hard to dispute Palmerston's assertion that England had, by mid century, become a nation made up of social classes; people whose prime and determining characteristic was social class. The fact that by 1851 England was viewed as an urban society and one dominated increasingly by social class was not coincidence. These two features of English life were new and interrelated.

Part Two

Problems and Solutions

The political response to the demographic and social changes in English urban life – especially legislative innovations – form the central core of much of the historical narrative of these years. New acts, however, were often inadequate to the tasks confronting them. The problems of poverty for instance were as endemic to urban life in 1851 as they had been in 1776, though by the mid nineteenth century they were much more extensive. Of course, it is easy to claim that the poor were (and are) always with us. What we can see in these years, however, is a transmutation in the very nature of poverty itself as the English became an urban people. Indeed, extensive poverty rapidly established itself as one of *the* characteristics of English urban life – and was to remain so thereafter.

To many contemporaries there were more troublesome problems than poverty. The English seemed to be changing into an ignorant and unbelieving people in direct proportion to the growth of the towns. How to tackle this problem became a source of political and social controversy; in the process the propertied perception of education for the lower orders was changed. At the beginning of our period popular education was widely distrusted and feared; at the end it was widely promoted. In these years, then, education for the common people changed, in the eyes of the propertied, from being a threat to, to being the salvation of English society.

One central reason for the distrust of popular learning was its traditional links with radicalism. And these were the years of a remarkable radical tradition; a powerful but changing movement of (primarily working) people determined to stake their claim to a wide range of rights. It would, indeed, be possible to write the history of these years through the eyes of the radicals. The purpose here, however, is to suggest the degree to which the developing radical movement was itself shaped by the changes in urban life. Radicals were distrusted and feared as a threat to law and peacefulness, not least because of the massive crowds they could muster in all phases of their campaigns.

The question of law and law enforcement became especially acute in the urban setting; towns and cities were apparently as volatile as they were medically dangerous. How to control the urban areas thus established itself as a key problem for national and local government. Always true of London, in these years it also became no less true of a multitude of other towns and cities. And yet, despite a host of violent incidents, English urban life was more peaceable by the mid nineteenth century than it had been at any period in living memory. Indeed, there is a case to be made that the development of a peaceable mentality – forces to the contrary notwithstanding – was one of *the* most remarkable changes in these years.

5 The poor

English history in the years between the onset of the American War (1776) and the collapse of Chartism (1848) is dominated by the most stressful of domestic social problems. English politics, high and low, were consequently troubled and often unpredictable. There were in fact few social groups in England who remained immune to these difficulties and uncertainties. It is hard to know which group felt more troubled; those plebeian groups buffeted and hammered by economic and social distress, or their governing betters who saw in the various political waves of popular unrest and resistance, threats of disturbance and revolt. We could in fact write the history of these years in terms of the English radical movement which so alarmed the entrenched governing orders. And however we view it, one of the key determining forces behind the radical urge, especially after 1791, was the ubiquity of urban distress, poverty and hunger.

There was perhaps nothing new in this. Hunger and poverty had, since time immemorial, haunted the common people. But what *was* new was the social context. Poverty and distress were more volatile and potent forces in a society characterized by population and urban growth. As towns grew in number and size those strata of urban life trapped in apparently inescapable poverty grew larger in numbers. They were to be found in huge concentrations throughout the length and breadth of the nation, and both in itself and in its manifold social ramifications the poverty of these English people posed one of the most serious, persistent and apparently irremovable problems of these years. Poverty as a major social ill became progressively more urban as society itself came to be dominated by towns and cities. Yet one of the major obstacles to coping with poverty was the residual outlook of the propertied classes who, by and large, continued to think of it and make proposals for coping with it in rural and traditional terms. Indeed, throughout much of our period, poverty continued to be handled through an ancient and creaking administrative structure.

Poverty had taxed successive forms of English government to varying degrees of intensity for centuries past. But it was the attempts of Tudor governments, notably under Henry VIII, to rationalize the system of poor relief that laid the basis for policy towards the poor which survived into these years. The late Elizabethan poor laws, which sanctioned local magistrates to appoint overseers, financed through local poor law rates, remained the political basis for dealing with poverty. Yet, beneath the apparently standard policy, there developed a multitude of administrative solutions. Whatever the local system, the basic problem – of poverty – worsened in the course of the eighteenth century because of the economic and demographic transformation. Population growth, social mobility and the drift to an urban and later industrial habitat, taxed the poor law system to the limits. And the problem was heightened by the economic crisis of the war years from 1793, more especially the ravages of hunger in 1795.[1] Moreover, in a society where perhaps the very great majority of the labouring population earned only marginally enough to keep a family in meagre circumstances, the slightest personal or industrial misfortune was enough to tip people into destitution. Some of the issues we have already examined, those 'unpredictable, yet inevitable financial disasters of childbirth, sickness and death' would instantly immiserate a working family. The plebeian family economy was always precariously balanced between tolerable sufficiency and dire poverty. And, as more people lived in towns, they faced that complexity of urban problems that could tip the balance the wrong way. Furthermore, this was all in addition to the inevitable ups and downs as a person got older, to say nothing of the economic or trade fluctuations which might, unaccountably, snap the working family's life-line to self-sufficiency.[2] The economic crises which regularly punctuated these years could, and did, plunge whole communities into distress. It has been calculated that 41 per cent of families in Oldham were below primary poverty in the crisis year of 1847.[3] And whenever a national – or more particularly local – economic crisis flared, destitution in the localities was intense, and with no clear or short-term solution in sight. Such problems – so frequent and regular throughout the nineteenth century – first began to make themselves felt in a major way in the crisis decade of the 1790s.

It had been widely accepted from Tudor times onwards that the 'impotent poor' – the aged, the young, sick and mad – were people who could do little, if anything, to fend off the ravages of poverty and were clearly and indisputably in need of material (and even institutional) assistance.

But the rapidly worsening – and spreading – of conditions of the 1790s, dominated by high prices and food shortages, focused attention on another group; those in work but unable to earn enough, even in full-time work, to prevent themselves and their dependants from sliding into poverty. From the 1790s through to 1914 (and of course beyond) it was the poor wage-earner who came to dominate successive investigations into the political solutions to, and the continuing problems of, English urban poverty. The Poor Law Commissioners of the 1840s, Booth in London and Rowntree in York in the 1890s all, and more, provided statistical substance to the central problem confronted by the Speenhamland magistrates in 1795; namely, that substantial sections of the poor were poor because their wages were too low.[4] The political and social crisis of 1795 served among other things to focus attention on the nation's poor. Parliament, however, failed to devise a solution that measured up to the problem or was national in scope, leaving, as before, the political initiative for handling poverty where it had traditionally rested – in the localities.

The most significant innovation came from the Speenhamland magistrates in Berkshire who decided to 'top up' local wages by a supplement that was regulated by the price of bread. This solution was copied in a number of places, but it would be wrong to claim that a 'Speenhamland system' evolved, although when faced by chronic and ubiquitous low wages, magistrates often opted to supplement them. However, the Speenhamland solution immediately incurred the intellectual and political antagonisms of contemporaries who viewed it as economically counter-productive (by undermining independence and self-reliance), as a depressant of wages and, perhaps most telling, it did not yield the social stability that the propertied orders believed to be the ultimate reason for paying their poor rates. As English society seemed to house swelling numbers of poor people, the country appeared to become increasingly unstable as the propertied orders found themselves pouring still more money into the local poor rates, they did not see a related peace and stability descend on their neighbourhood. This became a particularly acute problem after 1815, when the war was rapidly followed by worsening domestic conditions. The return of upwards of one-third of a million men from the armed forces, a downturn in economic demand, and national economic retrenchment cumulatively produced high levels of prices and unemployment. These, in turn, created a level of social and political turbulence unknown since the 1790s.[5] Not only did the urban world contain more – and more obvious – poor people (if only because of population rise) but the related changes in economic life, particularly the increasing economic specialization of certain towns and regions, meant that vast numbers of working people could be rendered destitute by a simple downturn in local economic activity. The consequent social volatility provides a descant to English politics from 1815 through to the end of our period. Politically, it was a relatively simple (and obvious) matter to fall back on to the repressive measures so well-oiled and practised throughout the war years to curb dissent and radical demands. What to do about the attendant and spreading levels of poverty was, however, less easily handled.

The poor laws were always an easy target. At the simplest of levels, they were ineffective. The poor were increasing, as were the poor rates, and yet there was no countervailing benefit of social calm. To make matters worse, the poor laws became the objects of withering criticisms by society's foremost economic and political theorists. Malthus, for example, dismissed them for allegedly encouraging the poor to have children before being economically self-sufficient. It was an argument compounded by Richardo's 'iron law of wages' which asserted that since there was only a limited amount of wealth for wages, the more money going into poor relief, the less into wages. To men in the localities, burdened by the poor rates and yet confronted by social unrest, such arguments, widely disseminated in a more popular and less subtle form, provided intellectual shape for their own feelings. And yet, however persuasive such arguments, the key question still remained; how best to modify or replace arrangements for handling the poor?

There were, it is true, a series of local experiments, notably in Nottinghamshire, involving a workshouse operated under such severe conditions that the poor law would be 'looked to with dread by our labouring class, and the reproach for being an inmate of it extend downwards from Father to Son. . .'.[6] Such a system (common from 1834) was successful in one respect at least; it became the most hated and feared object on the mid- and late-Victorian urban landscape. For the rest of the century only a fraction of the nation's poor resorted to the workhouse.[7] The political problem of poverty came to a head in 1830 during and after the devastation caused by the 'Swing Riots' throughout the rural southern and eastern counties. Although they were uniquely agricultural – characterized by rick-burning and machine-wrecking, and, despite their brutal though effective suppression, the riots raised troubling questions. Despite the £7 million spent each year on the poor law, it simply did not work, and sections of the labouring poor seemed to become violently desperate.[8] Suppression of the troubles was followed by an inquiry into the poor laws. Under the guidance of Nassau Senior and Edwin Chadwick, the Royal Commission on the Poor Laws, and its subsequent report, was to form the basis for the Poor Law Amendment Act of 1834. Both Chadwick and Senior concentrated their criticism – supported by an abundance of statistics – against the allowance system; of 'topping up' inadequate wages. Their argument was that it was a system which encouraged idleness and kept wages low. More recent research, however, suggests that such subsidies were generally implemented where local wages were already low.[9]

The purpose of Chadwick's scheme, swiftly implemented by Parliament, was to create a centralized system of poor law administration that was operated through the locally-regulated workhouse. It was decreed that within its bleak walls conditions were to prevail which were always *worse* than anything to be found outside – the principle of 'less eligibility'. Thus, the workhouse would cater for the destitute – the impotent poor recognized since Tudor times as unavoidably in need of care – but the rest (the poor as opposed to paupers) would be forced to care for themselves by recognizing that it was in their interests to remain among the ranks of working people. It was a social theory that made little sense when confronted by the new manifold problems of urban – especially industrial – poverty evolving in the Midlands and the north. The changing economic patterns of town life were, to put the matter simply, creating armies of people who were not, as Benthamites imagined, the authors of their own fate. In a society characterized by increasing economic specialization, where whole families and whole towns were employed in a specific form of economic activity, what alternative did impoverished working people have? The figures already given for Oldham in 1847 give the clue. Older towns – like London before – had traditionally provided a varied economic fare for labouring people who might in hard times, and if able-bodied, be able to turn to casual labour. This became progressively less possible in highly-specialized urban communities and for wide sectors of the labouring population.

Understandably, the 1834 Poor Law Amendment Act was bitterly denounced by radicals and working-class leaders, and, while its implementation in the rural south (at times of economic buoyancy) passed relatively unopposed, its arrival in the industrial north in the late 1830s (when times were again hard) was greeted by violence and resistance. To working-class spokesmen, the act was almost a caricature of their worst allegations about the exploitation of working people. From this passionate resistance there was born the Chartist movement, itself founded on the awareness that only political power could hope to deflect the worst excesses and indifference shown by the propertied to dispossessed working people. In the process there developed a bitter language of class.

Resistance to the new poor law was not uniquely plebeian. It was often opposed by local politicians and magistrates reluctant to lose their power to the London-based bureaucracy. Time and again, the local poor law guardians dragged their feet, refused, or were slow to follow instructions, sometimes ignoring completely the specifics or the general purpose of the 1834 act. The end result was that the implementation of the act was patchy and irregular, and the poor law administration survived in its essentially local and varied form. To make matters worse for the apologists for the act, the *cost* of poor relief actually increased, the workhouse was often slow in coming, and, even then, outdoor relief continued unabated. As late as 1854, 84 per cent of paupers were receiving outdoor relief.[10]

Statistics yielded by poor law relief seem on the surface to confirm the rising material standards of society at large. In 1834, there were some 1.26 million people on poor relief (8.8 per cent of the population). By 1850, this had fallen to 1 million (5.7 per cent). By 1880, it was only 808,000 (3 per cent). Yet, this evidence contrasts sharply with other data about poverty. Henry Mayhew, the founder of British empirical sociology, had begun to record and document the amazing depths of urban poverty, and, in 1848, he pointed out that, while 2.25 million (14 per cent) of the population had no gainful employment, a much *smaller* figure – 1.87 million – were calculated to be on poor relief.[11] It was clear from this and subsequent evidence that the poor law data provided no real guide to the true extent of poverty. The full extent of urban poverty was not laid bare until after our period, but even later in the century evidence from Booth and Rowntree, for instance, merely added substance to what many people already knew (and some could prove); that there were upwards of one-third of the urban population living on or below the poverty line. Yet the growing awareness of this figure simply added sharpness to what was visually obvious on the streets of urban England. Poverty haunted England to a degree and with an all-pervasiveness that became more offensively striking as society became more urbanized. It may well be that it was no worse, after a fashion, than seventeenth- or eighteenth-century London, but we need to recall that London had traditionally posed a major problem of how to cope with metropolitan poverty. What we can detect from the latter half of our study is the multiplication of London-type problems throughout urban England. Of course, the capital, because of its sheer size, was unique in the scale and diversity of its urban difficulties. But the governing orders were, by the 1830s, faced by a new, nationwide and evenly-spread problem of urban poverty. How to deal with the poor was in essence and in a particular form what to do about the governance and control of urban life itself. Furthermore, to suggest that poverty came to be seen as a particular *urban* problem is not to minimize the persistence of poverty throughout rural England.[12]

England in the early nineteenth century was a society increasingly obsessed by statistical analysis. Each and every contemporary problem – health, demography, poverty, industry, education – was swiftly rendered into statistical form by men anxious to know more about the exact shape of society at large. And just as they began to comprehend the full enormity of poverty through statistical evidence, so were they able to analyse its causes. In retrospect, many of the explanations may seem obvious. But, in tabulating the precise origins of poverty, it was possible for those who wished to do so to envisage and devise future political solutions. It was, for instance, clear by the 1840s that substantial proportions of the poor were poor because of low wages[13] – an ironic justification of the tactics adopted by the Speenhamland magistrates of 'topping up' insufficient earnings. It was equally clear by mid century that perhaps half of those receiving poor relief were sick or injured. But it was the old who were the most notable single group of people who slid into unavoidable poverty, though the exact statistical confirmation of this had to wait until late century. Whatever mathematical precision was bestowed on this and other causes of poverty, the extent – and causes – of poverty were widely appreciated by mid century. 'Low earnings, irregular employment, and large families, sickness, widowhood, and old age – these, rather than intemperance or idleness were the root causes of poverty'.[14]

At mid century, Henry Mayhew began his influential description of contemporary poverty, in a series of eighty-two pieces in the *Morning Chronicle* – some one million words in all, from October 1849 to December 1850. Their impact was as stunning as the evidence. Douglas Jerrold asked Mrs Cowden Clarke in February 1850: 'Do you devour those marvellous revelations of the inferno of misery, of wretchedness, that is smouldering under our feet?'[15] In large measure, the decision to write (and the *Morning Chronicle* to publish) a series on poverty had been shaped by convergent forces; relief at having been spared the ravages of the 1848 revolutions which had rocked Europe, relief at the fading away of the Chartist challenge, and, most critical of all, relief at the passing of the second attack of the cholera. It is important to remember that the social and political convulsions of both 1832 and 1848 were accompanied and accentuated by waves of apparently uncontrollable disease attacking urban England. And it seemed that both disease and political unrest were rooted and thriving in the squalid conditions of urban life. For those men of substance anxious to maintain their precarious political control over urban England, the joint attacks of disease and radicalism were symptomatic of the threat from urban life itself. It was therefore important to know more about the origins of that threat. It was from such unease that there was launched the *Morning Chronicle*'s seminal researches into the capital's poor. Henry Mayhew's work was the most instrumental, in shocking and shaping opinion. One man asserted: 'Those *Morning Chronicle* letters have set us grieving, thinking, and, I hope with some measure, acting.'[16]

What is striking, in all this, is the ignorance about the poor among their more prosperous contemporaries. Mayhew's articles appeared in book form in 1850 and his readers were widely amazed by his findings. In fact, Mayhew was fully aware of the public ignorance about poverty, describing the poor in his work as 'a large body of persons of whom the public had less knowledge than the most distant tribes of the earth'. One reviewer claimed that the work was 'full of facts entirely new'.[17] Such a reaction ought to caution the modern historian against overstating the degree to which poverty and the poor impressed themselves as a matter of course on contemporary political and propertied awareness. It took times of national or local crisis, or widely publicized propaganda, to jolt the propertied conscience (and even awareness) about poverty. Moreover, the shocks administered by Mayhew had to be delivered anew, at periodic intervals, by subsequent researchers. The English propertied public, despite the weight of accumulated evidence and continuing charitable efforts, readily and quickly forgot or overlooked the poor. There were periodic waves of social amnesia which seem to have swept through the middle and upper classes and which could only be countered by public assertions about the continuing existence of poverty. And this was *despite* the ubiquity of poverty throughout urban England. It may have been hard not to see or to notice the poor; but it was even easier to ignore or forget them.

It is easy, of course, to be critical of the better-off in not noticing the poor in their midst. Yet, it is crucial to grasp that their ignorance was itself a function of the nature of urban life. The residential 'zoning' of the different social classes (though differing from one town to another) served to isolate one class from another. This had, of course, been true of bigger cities in the eighteenth century. With the advance of urban growth it became ever more characteristic, as English cities developed their own particular modes of segregating the poor and the working class from the view, the ken, and the proximity of their social betters. Few English towns were quite so dramatically zoned as the seaside resorts,[18] but few were without their physical *cordon sanitaire* that isolated the propertied from the poor. Obviously, merely to travel around the town, or from one town to another, meant that the well-to-do were periodically confronted with some of the poorer aspects of life in contemporary England. But the concentration of poverty, the depths of its urban squalor, and the poverty in the form most actuely felt by its victims – in its domestic setting – was often hidden away from the casual view. It was because urban poverty was generally rooted in the often inaccessible recesses of urban life that the findings of the investigators, missionaries, and crusaders, were so important. Poor people could always be encountered in the streets of English towns, but their centre of gravity tended to be remote or masked by the physical contours of urban life.

No amount of self-help or personal determination could help the poor to break out of the poverty which ensnared them; often, like genetic family traits, passed from one generation to another. Indeed, it was an awareness of this helplessness (in conjunction with outraged feeling at the worst manifestations of poverty) that provided inspiration for the growing tide of nineteenth-century philanthropy. Yet, for all the unstinting and exhausting charitable efforts (often by women), and notwithstanding their success, it was a mere 'holding operation' that was forever in danger of being overwhelmed by the continuing magnitude of the problems of poverty. By mid century, there were people who had come to realize that English poverty demanded more than mere charity. The political response to urban poverty was to become a recurring and ever more voluble descant in radical (later socialist) politics for the rest of the century.

6 *Learning and believing*

The physical changes in England in these years are easily described because they were so visible. Even the poor – however subterranean they may have seemed to many early Victorians – were physical landmarks in the urban landscape. Less obvióus, though no less important, were the changes in attitude brought about by the inexorable drift towards urban – in part industrial – life. To a marked degree this was the case, as we have seen, with the question of class consciousness. And there were other marked changes; notable breaks with older attitudes and values. In the course of these years there was, for instance, a quite dramatic change in attitudes to learning and education – especially for the common people. In the mid eighteenth century learning (however rudimentary) among labouring people was generally viewed with suspicion and dislike by their betters. Frequently denounced as subversive and dangerous, popular learning was positively resisted. By the end of our period, however, quite the opposite outlook prevailed among the propertied orders who now believed that ignorance, not learning, was the greatest threat to and the solvent of social harmony and tranquility. And, as if to confirm their fears, it was revealed, in irrefutable detail in 1851, that religious belief (long viewed as a major force in securing loyalty and stability) was itself in decline.

Formal learning had for centuries past normally been in the hands of clerical bodies, or men whose training and direction were clerically inspired. In more prosperous homes, literate mothers had long been instrumental in initiating and encouraging literacy among their offspring, both male and female. Yet, access to the printed word was arguably as important in maintaining and furthering literate interests as the acquisition of basic literacy. The problems of maintaining literacy often compounded those of acquiring it. And while it might seem straightforward that literacy was of obvious utilitarian and social use in the upper and middle reaches of society, it is less clear what benefits poorer groups might derive from it. Of course, it is very easy for modern historians to overvalue the importance of literacy. We need only look to poor countries today to see how millions of people survive without literacy. Non-literates today, as in early modern England, were often informed by their literate peers. Reading aloud, word of mouth, folklore, and a sharp, but unrefined, alert intelligence, all ensured that the non-literate nation was kept abreast of news and events. In this process, few places were more seminal than the ale-house, for it was there that news or gossip, information and rumour passed to the transient public, often via the contemporary press. It would then be quite wrong to assume that learning, in its broadest sense, was the same as literacy.

Assessing levels of literacy is doubly difficult because of the problems of sources, for, while there is an abundance of material (the outpourings of literature of all kinds since the Tudor period), the records of formal school instruction and of informal (generally domestic) instruction, such material does not easily yield clear statistical patterns of literacy. Fortunately, we are helped by the fact that, from 1754, husbands and wives were expected to sign the marriage register. Notwithstanding the enormous difficulties in assessing such data, historians have been able to make fairly confident assessments about literacy for the early part of our period. By about 1760, evidence from Yorkshire parishes, for instance, suggests that many more men were literate than women (64–39 per cent), although there were very great variations from place to place. Equally, there seems to have been a general improvement in overall literacy in the second half of the century, though this cannot always be explained by the presence of a school.[1] The pattern in London was even more impressive; in much the same period some 92 per cent of men and 74 per cent of women had minimal literacy. In provincial towns, literacy was higher than in the surrounding countryside – but lower than in the capital.[2] There was, at the opening of our period, firm evidence of the relationship which was to become clearer with the passing of time; urban life in general encouraged greater literacy. There were a number of reasons for this. The nature of economic life in urban areas demanded greater literacy as an employment skill, and towns were of course centres of printing and publishing. In addition, formal instruction was more easily organized in urban than in rural settings. For these reasons alone – and there were many others – it might be felt that literacy would increase as society became more urbanized. The evidence, however, is less clear cut. By the time of the 1851 census, literacy had made painfully slow progress in many urban areas. It is to be expected perhaps that mid nineteenth-century rural illiteracy would remain especially high. But it was little better in many towns and cities. Throughout the 1840s, for instance, evidence from a variety of sources continued to paint a bleak picture of the levels of illiteracy. This was particularly true of very poor urban girls. Among men, the picture was brighter; some 66 per cent signed registers in 1840, and the percentage increased steadily (by 1870 it stood at 80 per cent).[3]

What is perhaps remarkable is the consistency of male literacy. In the fifty years from 1754, it stood at about 60 per cent; it began to rise steadily from about 1805 onwards. For women, it remained at less than 40 per cent in 1754 and a little over 50 per cent in 1840. Of course, these *national* figures mask a host of regional, local and even parish variations. Rural society traditionally lagged behind the towns. But the towns themselves had remarkably different degrees of literacy. Older market towns were well in advance of the new industrial areas in the nineteenth century. Up to the end of our period, the impact of industrialization seems to have had a harmful – sometimes a disastrous – effect on local literacy. In Ashton-under-Lyne, it fell from 48 per cent to 9 per cent between 1823 and 1843. But we need to remember that the factory was not always instrumental in this decline. An equally crucial force was the sheer pressure of population on threadbare educational institutions and the continuing survival of outwork that depended heavily on child labour.[4] Moreover, even when formal education was provided, notably in factories, the results were sometimes appalling. Only 199 of 963 unemployed girls at a sewing school in Manchester in the 1860s, who had attended school, could read and write.[5] It seems reasonable that industrial change, especially the textile factories and the expansion of outwork, both of which relied extensively on child and young adult labour, retarded the educational development for large numbers of the nation's young. But this point also serves to illustrate the diversity of urban experience. Whereas towns had traditionally been in the vanguard of popular literacy, there were a number of early nineteenth-century towns which lagged well behind the low literacy rates so characteristic of country life.

Literacy, however, was much more than an educational issue, and the bare statistics mask the social transformation through which literacy became a major political issue. Throughout the years 1776 to 1800, and again after 1815, England was shaken by political and social movements which were themselves deeply influenced by economic changes. The American war, the revolutionary and Napoleonic wars, followed by years of post-war dislocation, were paralleled by domestic political turbulence. Moreover, it is important to recall that the local radical movements spawned in these years were firmly committed to the use and furtherance of the printed word.

The strength and appeal of radical movements – from Wilkes to the Chartists – were to be found at the lower reaches of society. To the startled outsider, notably those in government who were determined to resist the progress of radicalism, there seemed to be a close, if not causal, connection between popular literacy and radicalism. By the years of revolution, there seemed ample evidence for feeling that those humbler sorts who could read would never be happy with their lot in life.

The connection between radicalism and widespread literacy can be traced in the short term to the Wilkes movement which spawned a rich and varied literature and engaged a 'political nation' far beyond the narrow confines of propertied society. In the words of a pamphlet of 1765, there was, in England, 'scare a human creature so poor that it cannot afford to buy or hire a Paper or a Pamphlet, or so busy that it cannot find leisure to read it'. Nor was this feeling merely a partisan political claim for, despite the obvious and manifest problems associated with defining and interpreting literacy, there was, in the later eighteenth century, a veritable deluge of printed material which tumbled from the presses. Newspapers grew in numbers and copies in London and the provinces; the 7.3 million sales of 1750 had swollen to 12.6 million in 1775. In London there were four dailies and nine thrice-weekly papers – quite apart from lesser ephemera. A similarly healthy picture existed in the provinces where the thirty-five to thirty-seven papers of 1760 had grown to fifty by 1782. But this was just the tip of the publishing iceberg, for a plethora of pamphlets, tracts, and abstracts fed the growing appetite for terse political literature – all, of course, in addition to an abundance of printed ephemera, songs, handbills, and squibs, which often sold in tens of thousands.[6] Naturally, these and other printed materials grew in volume – and popularity – in times of political excitement and tension, feeding the reading public through the coffee and ale-houses and hawked in the public place. Thus, long before the transforming impact of revolutionary ideology, personified by Tom Paine, there was a rich and vibrant popular literacy and literature that was closely associated with, and propagated still further by, radical politics. In 1776, it was well-established that widespread literacy was a basic feature of urban life and that it had an intimate and symbiotic association with the progress of radical and reforming politics.

To those in political power, in Westminister or the localities, this association became ever more marked, troublesome and dangerous as the eighteenth century advanced. It was, for instance, a characteristic of the radical agitation spawned by the American revolution – the Association movement and the Society for Constitutional Information – that although their membership was socially exclusive, their tactics of publishing and disseminating cheap or free tracts enabled them to reach a much wider and humbler audience. And this was, of course, in addition to events in the American colonies where the association of widespread literacy, far-ranging cheap publications, and easily accessible libraries, was of crucial importance in forging the urge for independence. In Britain, this tradition was revived after 1787 by the formation of the Abolition Society to end the slave trade, and keen to harness public opinion to that end. But it was the impact of the French Revolution which utterly transformed the phenomenon.

After 1792 and the irrefutable impact of Paine's writings among tens of thousands of working men throughout urban Britain, it was impossible to deny the link between the cheap printed word and the radical movement. Paine quickly became the prophet and scribe of an English revolution which, if not openly violent, was none the less disruptive in the eyes of its propertied opponents. Bitter complaints in the 1790s plot the trajectory of Paine's sales. From every part of the country – including remote rural areas, but particularly from the towns – men of substance complained of the sales and availability of cheap and abbreviated copies of the *Rights of Man*. In the Potteries 'Paineite publications are in the hands of most of the people . . . particularly the journey men.' In Manchester, Paine's ideas were affecting 'the minds of the lowest order of the people'. What alarmed Pitt's government and its supporters throughout the country was the rapid accumulating radical sentiment among men of a humble station, and the fact that their arguments were rational, articulate, and literate. Indeed, it is symptomatic that among the government's counter-measures (which included discriminatory laws, local intimidation and arrest) was the establishment of a barrage of alternative literature to counter the Paineite tradition. By 1793, the government had come to terms with the existence of a mass reading public; the problem remained, however, what to do about it ? Hazlitt later wrote:

> When it was impossible to prevent our reading something, the fear of the progress of knowledge and a *Reading Public* . . . made the Church and State . . . anxious to provide us with the sort of food for our stomachs which they thought best.[7]

Thus, by the early nineteenth century, there had clearly developed a political view of popular literacy and education; a view and an assumption that it was closely associated with radical politics. Throughout the French Wars, in the subsequent years of post-war political troubles, and in the heightened radical agitation of the 1830s and 1840s, different governments invariably displayed a marked hostility to radical and popular publications.

Yet, the public demand for cheap literature was manifest, a fact underlined after 1816 by William Cobbett's launching of the *Weekly Political Register* and his determination to make his radical critique available to a wide audience (the Stamp Act prevented it). Throughout the 1820s and 1830s, radical and working-class organizations and publications bitterly attacked the Stamp Act – the 'tax on knowledge' – viewing it as a deterrent to the spread of plebeian learning and knowledge, and thus blunting the edge of radical ambitions. Radicals and their opponents were oddly in agreement on one point; to use the words of the *Poor Man's Guardian*, 'knowledge is power' – and, for that reason, it was an article of radical belief – and propertied opponents resisted it accordingly. And it was not until the tax was first lowered (1836) and then abolished (1855) that there was released a flood of cheap publications. Throughout much of this period then, each radical phenomenon – the American war, the French Revolution, post-war dislocation, and, perhaps most notable, Chartism – was characterized by, among other things, the development of plebeian political organizations and publications. Indeed, the radical case, whether it be reform of Parliament, abolition, trade union reform, Catholic emancipation or factory agitation, was advanced by cheap and easily accessible publications. Consequently, it was a feature of the opposition to radical organizations that it should include hostility towards the cheap printed word. It was assumed that widespread popular literacy, fed by the scribes and publishers of radical literature, was socially disruptive; that the fabric of political society was weakened by the critical scrutiny of popular literacy.

Yet, later in the nineteenth century an utterly contrary view prevailed (though one whose genesis can be detected in these years), namely, that ignorance and illiteracy were socially harmful. Indeed, it was to be one of the marked transformations of the nineteenth century that literacy and widespread education were to be seen as the cement, and not the solvent, of social cohesion and stability.

Radical organizations were not, of course, the only agencies for promoting or channelling popular literacy. The late eighteenth and early nineteenth centuries were remarkable for the educational innovations throughout the country. Elementary schools proliferated, notably in the towns, as churches, voluntary agencies, and charitably-minded individuals, struggled to bring a little learning to what they saw as an ignorant and godless common people. The swarms of ignorant children which slowly but surely came to dominate the urban landscape demanded attention. And from the determination 'to raise the lower classes in habits of industry and piety' there developed after 1787 the Sunday schools whose founders, Robert Raikes and Hannah More, were appalled by the ignorance and the godlessness of poor children. There was, inevitably, fierce resistance to the education of the poor. In the words of one farmer's wife: 'the lower orders were fated to be poor and ignorant and wicked: and that as wise as we were, we could not alter what was decreed'.[8]

The urge for learning was shared by many working people themselves – as the history of the working-class movements confirms. The convergence of these varied demands for basic education – though often inspired by conflicting religious interests – produced a transformation in early nineteenth-century schooling. By 1818, there were some 630,000 children on school registers (6.5 per cent of the population); by 1851, it had grown to 2 million (13 per cent). In the Sunday schools – arguably the most important pioneers of bringing education to working people – the 450,000 scholars of 1818 had multiplied to 2¼ million by mid century.[9] Moreover, this remarkable expansion was accompanied by a shift in the nature of the schools themselves. In the mid eighteenth century only 20–30 per cent of pupils attended schools controlled by a public body; most attended voluntary and private schools or were tutored. A century later, 68 per cent attended publicly-controlled schools – a trend, of course, ultimately confirmed by the coming of compulsory education in the last quarter of the century.[10] And, as the numbers of pupils grew – most notably, of course, in the towns – the small schools, the general absence of formal discipline and curriculum – all began to give way to the need for greater organization, rationalization, and control. This process was most clearly at work – and best recorded – within that small band of ancient public schools, but a similar process (influenced by similar social pressures) can be seen within schools catering for poor children.

It has been customary for historians of the growth of English education to describe this transformation in terms of the organizers and initiators of education; to view it as a creation of the middle- and upper-class educators anxious to provide working-class people and their children with what they thought was best for them. Thomas Laqueur, on the other hand, has provided a healthy reminder that working people were not merely passive victims in all this; they were not, as many historians would have us believe, the mere potter's clay from which their social superiors were able to mould the outlines of malleable and controllable social order. There were many thousands of working people, from one generation to another, who wanted and secured a little learning for themselves and their offspring – a fact amply illustrated by levels of literacy long before and during this revolution in popular educational provision. Moreover, it was a matter of continuing puzzlement to interested propertied contemporaries that working people often chose to send their children to more expensive schools (local dames' schools, for instance) in preference to cheaper local schools. In Warrington in 1837, in common with large numbers of other industrial areas, it was concluded that:

> a sense of duty and interest induces a large proportion of the working class to make considerable pecuniary sacrifices for the mental cultivation of their offspring.[11]

But they did not send their children to the local church school. Throughout industrial England in the early nineteenth century, the working-class consumers of education exercised their choice with fine discrimination – ignoring the schools with poor teaching, patronizing those with a good reputation.

It is dangerously easy to romanticize this phenomenon – or to make excessive generalizations or conclusions from it. We need to recall that the education of working-class children was always dependent on overriding and determining economic considerations. The family economy – and the vital role within it played by children – often demanded a premature end to schooling. Time and again, working-class autobiographies described the ending of school days in economic terms:

> The income of my parents rendered it necessary that I should do something for my living as soon as possible, and accordingly I was sent to work before I was eleven years of age.

Another man wrote:

> Squire Clough kindly sent brother John and me to school at Thirsk; but we only attended for a short time, when my father left us, and my mother, though toiling hard to keep the home together, had to avail herself of any small labours we elder boys could render.[12]

Not only was elementary education formally concluded, but it was often punctuated by passing or temporary exigencies; sickness, unemployment, family misfortune, and, more especially in the rural world, the dictates of agricultural life. One boy left his school at 8 to care for 'several younger brothers and sisters', returning to school a year later. There was what David Vincent has called the continual interaction between poverty and attendance at school. Indeed, in his important study of early working-class autobiographers, Vincent makes the telling point that, despite the obvious fact that the autobiographers were exceptional people – a tiny and impressive élite – they derived only the most rudimentary of learning from their schooling. On numerous occasions, this band of plebeian intellectuals stressed the inadequacy of their formal learning when they were finally obliged to quit school. Only three could claim more than a basic acquaintance with writing and arithmetic. One man wrote angrily, in a judgement that covered the experience and fate of millions,

> Our poverty has been an insuperable bar to admission into any school, save the so-called charity schools, and a mere apology for learning doled out to us – just allowed to learn enough to 'get the catechism', some to write their own names, few to learn the simple elements of accounts – then, with such acquirements, early in life, started to the factory or dung yard, to earn a few pence to assist in procuring a family loaf of bread.[15]

Of course, it could legitimately be countered that such men by their very achievements felt their educational and social achievements all the more acutely; their frustrated talents forming the cutting edge of such cynical but truthful commentary. Throughout these years of social change, there was an undeniable and often highly effective working-class demand for elementary education. But, like almost every other characteristic of working-class life, it was determined, shaped and ultimately fenced in by the harsh realities of working-class economic life. Elementary schools, at best,

> remained a fragmentary experience, subordinated to the demands of the family economy and unlikely to endow even the most privileged and intelligent of working class children with any more than Arch's 'rudiments' of literacy and numeracy.[14]

In the process, perhaps the *key* factor was not so much the type of school attended, but working-class parents. It was they who paid for – or could not afford – the school fees, chose the school, encouraged rudimentary learning, and generally determined their children's commitment to learning. As with many other social characteristics, the initial – and continuing – exposure to elementary learning was initiated, paid for, and insisted upon, by parents. Generally, this was at considerable cost to themselves and their families, especially at times when money was scarce and the economic future uncertain. Indeed, the expanding volume of working-class elementary education throughout these years is a firm indication of the persistent working-class parental belief in education for their young, however inadequate it might seem, and however limited the social and economic opportunities it might afford in the future. However cynical the modern historian might feel about the purpose and value of that education, it was a cynicism not altogether shared by many working people in the early nineteenth century, who made such sacrifices to secure for their offspring a rudimentary learning.

In this process of ever-widening popular literacy, perhaps *the* most crucial institution was the Sunday school (an institution which became a major feature of urban life). The ambition of the evangelical founders of the Sunday school movement was to create an educational instrument that would control and pacify the nation through the inculcation of good habits of piety and hard work. Yet, the success of those schools cannot be judged solely by the aims of their founders, but rather by their development within those working-class communities of the nineteenth century which proved so fertile a seed-bed for their growth. The thousands of Sunday schools, and the millions of children who passed through them, were instrumental in encouraging an appetite for the printed word which could not be satisfied solely by the sectarian publications of those schools, or by their larger church organizations. Paradoxically, it was the awareness that, once established, popular literacy could not be staunched, which persuaded conservative evangelicals to establish Sunday schools in the hope of directing plebeian literacy into the correct (i.e. the loyal and counter-radical) channels. The end result; however, was often quite different.

The Sunday schools and the popular literacy they enhanced fed into an autodidactic working-class culture which itself became an agent of plebeian radicalism – and which continued well into the twentieth century. It has become a truism to claim that the rise of the twentieth-century Labour Party owed more to Methodism than Marxism; but it is a truism which represents a significant and often distorted historical substance. The organizations of working men – the trade unions, the Chartists and others – were characterized by those qualities which their members had often acquired and developed in their younger days in the Sunday schools. The epicentres of popular radicalism in the first half of the nineteenth century were those same towns where Sunday school activity was most pronounced, and, time and again, men came to a radical awareness via the rudimentary education and literacy provided within the Sunday schools for their childhood. The more general proposition, so frequently and glibly advanced, that religion was a counter-revolutionary force, needs careful scrutiny. Indeed, it is revealing to know that, at precisely that same time, the development of the non-conformist churches among the slaves in the West Indies and the US South (from the same sects flourishing in urban and industrial Britain) had a very similar effect; bringing organization, self-help, autodidactism and radical leadership, to the ranks of the slaves. Slave owners no less than the propertied conservative ranks in England feared the advance of plebeian organization and literacy – and for very similar reasons; that both would prove the solvent of their insecure social grip.

Of course, it is perfectly clear that many of the results of popular religion – in this case, of the Sunday schools – were unintended by the founders. For many, the lessons of Sunday schools were meagre, ill-learned and often of little consequence. For others, however, the Sunday schools were the first step towards a more rounded and complete education. Ben Brierley, the Lancastrian working-class radical, recalled:

> Aspiring to know more than could be taught me at the Sunday school in Hollinwood, I joined the one known as the 'Old School', in Pole Lane, Failsworth . . . there I found a large number of congenial spirits, who, like myself, had grown out of their childhood, and were looking forward to becoming men. We banded ourselves together, and formed the nucleus of the present Mechanics Institute, then existing under the name of the Mutual Improvement Society.[15]

While it may be argued that such men were few and far between, the ranks of the Sunday school scholars – from whence these men sprang – were massive. By 1851, there were 2 million children attending Sunday schools. By the end of our period, the Sunday schools had become major landmarks in the geography of urban England and a seminal force in diffusing the manifest personal and collective consequences of popular literacy throughout plebeian society. The most recent historian of those schools has argued that in these years there took place 'the birth of a working class culture that was deeply rooted in that ethic of education, religion and respectability which was embodied in the Sunday school'.[16] A number of the salient features of nineteenth-century working-class life were, then, shaped by Sunday schools. There is a distinct irony here, for the inspiration behind their founders had been primarily rural and pre-industrial. In the event, their most lasting influence was to be among the working class of industrial England. Whatever the ambition of their founders – to inspire an unquestioning and religious loyalty – the Sunday schools were to spawn many of the men, attitudes and organizations which posed the most serious challenges to the status quo from the 1830s onwards.

Despite the millions of children in Sunday schools in the mid nineteenth century, it was a matter of great concern to many pious Englishmen that the nation was losing its religious commitments in proportion to the increase in urbanization. The close links between religion and learning were traditional. The Protestant Reformation had confirmed the feeling that formal instruction was the right and proper preserve of the church and of churchmen. Indeed, from the sixteenth century onwards there had been a spate of new schools and colleges designed to 'advance the Protestant Reformation by banishing ignorance and implanting knowledge of the truth'.[17] By the late eighteenth century, the Protestant Church of England was the unquestioned and dominant religious force. But its role was enfeebled by the social and economic forces that were rapidly transforming English society. In the face of demographic change and encroaching urbanization 'the machinery of the religious Establishment broke down'. The Church of England found itself structurally incapable of responding to the early phases of this change. The Church's pre-industrial structure was designed especially to minister to the south, the south Midlands and the south-east, with a concentration of parochial effort devoted to rural life where it could work effectively in conjunction with the 'landed interest'. It was 'best adapted to those regions, occupational groups, and cultural contexts which contained the bulk of the population'.[18] As the demographic and occupational nature of society itself was slowly but inexorably transformed, the Church found itself stranded; ever more distant in time, space, and spirit, from the urban heartlands in the Midlands and the north. The Church of England simply could not respond to the initial urban and human transformation of English society. In the years when the population grew by almost 10 millions (in the century to 1841), the Church of England registered little increase in its manpower, or little readjustment of its parishes to accommodate the expanding population. The shift in the balance of the English population had effectively removed the people from the supervision of the Church. Moreover, that supervision was, by any standards, increasingly inadequate and weak.

The established Church had long been renowned for the vices of pluralism and absenteeism, and general indifference towards pastoral care by its churchmen. In sharp contrast, there developed an expansive, free-moving, preaching non-conformity, able to outflank the Church of England and its parish boundaries through its peripatetic ministers. The preachers on horseback took their message into the heart of the new urban and industrial settlements. Furthermore, the people at the centre of those settlements were precisely those most vulnerable to the blandishments and seductions of the new dissenting message. Long before the noticeable advance of urbanization, non-conformists had been successful in winning to their side those artisans, outworkers and traders, and others not closely tied by deference or economic dependency to local landowners – and thus indirectly affiliated to the local Church of England.[19] Those individuals and communities more securely tied to the land by various obligations were much more likely to retain a formal allegiance to the values and controls of the established Church. Conversely, non-agrarian groups were those most susceptible to the non-conformist appeal. It was then not surprising that in the initial phase of industrial growth (to the 1840s) in those early industrial villages of the Midlands and the north which formed the skeletal framework for industrial change and development, the non-conformists found a receptive and yielding flock. In coal and iron communities in Northumberland and Durham, in the textile districts of Lancashire, Yorkshire, Derbyshire and Nottinghamshire, the Potteries of Staffordshire, the growing population gathered together in industrial villages and small-scale factory and mining communities. Up to the 1840s, such areas were characterized 'by the survival of capitalistic outwork as the predominant form of industrial organisation'. And, at the heart of that outwork, lay an expansive labour force of artisans and domestic workers who formed the first labour force for the initial development of the industrial revolution. While it is true that they went into decline after the French wars, in the early part of this study they greatly expanded. Hand-loom weavers, for example, increased from 50,000 (1769) to 240,000 (1820). They and other artisans and outworkers formed the human base for the massive expansion of non-conformity. It was among such people that Wesley and his followers made conscious and effective inroads. Thus, the early phase of industrialization helped to hasten the breakdown of the old parochial system and this breakdown was consciously and effectively exploited by Wesley and the Methodists. Methodism seemed at its strongest where Anglicanism was at its weakest. And when, in 1828 and 1829, Parliament removed the disabilities against dissenters and Catholics, it seemed to many that the former hegemony of the established Church was threatened by imminent destruction. Only in the early 1830s did the established Church begin to awake from its slumbers.[20]

The true nature of early and mid nineteenth-century religious practices was revealed in the Religious Census conducted on 30 March 1851. The Census had its methodological flaws and its statistical weaknesses, but the broad sweep of its conclusions have generally been accepted as an accurate guide to contemporary religion. While the population of England and Wales was 17,927,609, only 7,261,032 attended church services. Even allowing for the young, the sick and the old, it was clear that 5,288,294 could have attended church, but failed to.[21] Today, such proportions would be offered as evidence of widespread religiosity, but to the early Victorian it provided irrefutable proof of the decline of faith and the widespread attack on religion. Even more alarming, to those who viewed the Church of England as the spiritual pillar of the state, was the fact that only slightly more than half the churchgoers attended the established Church. Not surprisingly, dissenters made great play of this fact. What right did Anglicans have to claim to be 'the National Church'?

The evidence provides an abundance of insights into contemporary religion at mid century. Church attendance was higher in rural areas and small towns than in towns with populations of 10,000 and more. If, as was clearly the case, England's population was rapidly concentrating itself in bigger towns and cities the long-term prospects for the established Church were poor indeed. Moreover, the absentees were primarily the urban working class. It now seemed clear (though concerned churchmen had long been aware of it) that the churches were predominantly middle-class institutions. The high Tory values of the Anglican clergy were not only anachronistic in themselves, but were also in sharp conflict with the urban and industrial culture of ordinary people.[22]

This is not to deny, however, the major and influential reforms set in train from the mid 1830s by worried Anglicans, whose own experiences and observations had long predicted the religious difficulties later documented by the 1851 Census. Indeed, by the time of that census, there was a new breed of Anglican activist – clergymen and laymen – fired by a new commitment and activity and at work in the poorest quarters of working-class life. If it is true that the overwhelming majority of English marriages and funerals were according to religious rites, and that large numbers of people continued to celebrate the major Christian festivals – Christmas, Easter, harvest festivals and the like – it was indisputable that formal belief was not a dominant social characteristic of the urban working class. And yet many working people, who themselves failed to attend church on a regular basis, dispatched their offspring to the Sunday schools. Of course, in crowded homes – in the days before compulsory schooling – there may be a simple explanation for this; with children at Sunday school, parents were allowed a brief and precious time to themselves, with the added bonus of peace and quiet and the possibilities of uninterrupted sexual activity.

There were also unmistakable strands of deism and unbelief in early Victorian England, though the morality of the formal secular movement, centred on the work of George Jacob Holyoaks, was as severe and stark as that of many churches. Notwithstanding the unbroken roots of free thought, stretching back to the influence of Tom Paine in the 1790s, the major impact of organized free thought was to be felt in the 1850s and beyond.[23] People were not lured away from the churches by free thought but by that complexity of social and economic factors that were, quite simply, eroding and transmuting the role and nature of organized English religions. Indeed, this point was made by Horace Mann, the organizer of the Religious Census. Working people were, he thought, 'unconscious Secularists', hostile or indifferent to formal religion for social reasons; they suspected prosperous ministers, they were poor, and distrusted the class system manifestly at work within the Church. Henry Mayhew illustrated the point – as he did so many other issues – with glaring detail, and in the words of working people. One told him:

> Religion is a regular puzzle to costers. They see people come out of church and chapel, and as they're mostly well-dressed, and there's very few of their own sort among the church-goers, the costers somehow mix up being religious with being respectable, and they have a queer sort of feeling about it. It's a mystery to them.[24]

There was nothing mysterious, however, about working-class Catholicism. And for that, we need only turn to the massive immigration of Irish. Between 1793 and 1815, 4000 landed each year in England, a modest influx which continued through to the 1820s and 1830s. In the 1840s, however, in the wake of the famine, some 40,000 Irish immigrants arrived *each year*. Between June and August 1841 an astonishing 60,000 left Ireland for the mainland. Indeed, between 1841 and 1850, Irish immigration accounted for a little under *one-fifth* of the total increase in the mainland population. And they brought with them their beliefs and their priests. By mid century, it had become clear that the Irish settlements in England had made it possible for Catholicism to become, as never before since the Reformation, a genuinely popular religion.[25] The conversion of England was to remain a dream – but not an apparently unrealistic one – within the Catholic hierarchy throughout the second half of the century. In this, as with other features of English religious history, the experience of the first half of the century was to be used as a basis for change and development in subsequent decades. In some respects, this also happened with Methodism, which, notably in rural society, provided a more attractive religious alternative for working people.[26]

What seems clear is that these generalizations about English religions need to be set against a mosaic of local variations. The crude distinctions between rural and urban life distort the multiplicity of religious experiences which often varied from town to town.[27] Religious forces, the different local experiences of a particular social class, the size of a town, the nature of local industry, the ratio of local inhabitants who had migrated from rural parts, created distinctive local religious patterns which marked off one town from another. In York, for instance, with its myriad of ancient parishes and its economy relatively untouched by the machine, the Church of England, and a number of other sects, were quite successful in reaching large numbers of working people. Indeed, it has been calculated that there was an attendance rate of 53 per cent at the census in 1851.[28] Such a figure, set against the low attendances for other communities, illustrates the great variations and local differences. In Bethnal Green, on the other hand, only 6024 from a population of 90,193, attended church, while, in the industrial north (on a line between Gloucester and Grimsby) only six of thirty-seven towns managed to reach the average church attendance of 58 per cent. And this is, as we might expect, dramatically true even within the same city. In an urban society that became increasingly divided by social class, confirmed by geographical zoning and distancing of one group from another, it was pointless, as Hugh McLeod has shown, to expect people from different classes to meet in the same church.[29]

It is reasonably understandable that the wretchedly poor and distressed seemed beyond the pale of formal and organized religion in these years. Despite a number of valiant local efforts, it was to be much later in the century that churchmen (and women) began to address themselves directly to the problem of the urban poor, through settlements and other missionary activity. The apparent inability of many religious groups to make substantial inroads among the working class in the major cities was a recurring source of complaint. A congregationalist recorded:

> it is a matter of deep regret, and not less of astonishment . . . that the artisans and working men of England are so rarely drawn within our circles.[30]

While it may be tempting to imagine that religion offered the very poor and the dispossessed a way out of their troubles by dangling before them the prospects of future happiness or salvation to come, in fact the opposite seems to have been the case. The destitute seem to have been 'too absorbed in the struggle to remain alive to look for "other worldly compensations", and, if they thought of the Creator at all, they were likely to blame Him for their sufferings'.[31]

There are, of course, striking examples to the contrary. This was especially the case with the strength of Catholicism among very poor urban workers, particularly the Irish immigrants, where the church attenders were often faced by the problem of not having enough physical space to worship. Similarly, a powerful plebeian commitment to local religion was to be found in the smaller industrial towns (or, better still, villages) and in certain rural areas. But, in these cases, the particular religion was able to identify itself with the wider community – and came to play a crucial and determining role in the cultural life of that community.

The census on the state of religion in 1851 provided more than an analysis of the state of contemporary religiosity, for it also laid the basis for future change. Though the evidence itself became a source of religious and political dispute, it none the less enabled interested parties to address themselves to the pressing difficulties.faced by organized religions. What was done by the various religious organizations lies outside this study – and each denomination's response was different. It was clear to all, however, that substantial pockets of the population were uninstructed – in religious or secular terms. They had no access to, and remained utterly ignorant of, formal instruction of any kind, and the evidence from 1851 confirmed long and widely-known truths about the need to bring education to the poor. It was equally obvious that those churches responsible for education were failing in their efforts; vast numbers of the urban working class were not only ignorant and uneducated but were indifferent or hostile to religion itself. How best to make good these startling shortcomings become an issue of sectarian and political contention throughout the rest of the nineteenth century. The coming of compulsory education was ultimately to settle the matter of who should rectify the nation's collective ignorance and irreligion.

The 1851 Census did more than this, for it was grist to the mill of those who argued that ignorance – in all its forms – was more dangerous to the fabric of social order and well-being than the presence of a little learning. 1851, then, represents much more than an arbitrary turning point, for it provides a historical landmark. It was a chronological point at which it had become clear (to those willing to listen) that English society, for all its encroaching material well-being and propertied pride in the initial achievements in commerce and industry, was fissured not merely by the scars of social class but by massive urban patches of ignorance and irreligion. For a society which prided itself in its learning, achievements, and its unparalleled industrial progress, it was disconcerting, to say the least, to learn of such massive social problems. But it was equally clear that the shortcomings of piety and learning were not isolated phenomena to be tackled and solved within themselves. They lay rooted in the confusing difficulties posed by urban poverty and deprivation. And the solutions to those problems were to remain a political and social preoccupation throughout the succeeding decades.

Untold thousands of English people were irreligious and ignorant, because they were poor, and although this formulation is a crude one it was to remain basically unchallenged well into the twentieth century. And what made their poverty and ignorance so difficult to solve was that both were a function of the distinctive deprivation of urban life. More than anything else, what was needed was amelioration of the urban environment.

7 *Challenging the system: radicalism and the development of mass politics*

It is tempting to imagine that England in these years changed inexorably under the pressure of forces – economic and demographic – that no one could control or direct. There were, however, powerful efforts consciously to change society, to arrest its particular line of development, or to direct it into new (and sometimes old) paths. There had been, of course, an alternative (though varied) image for the future of English society since the seventeenth century. Indeed, many of the radical ideas spawned by the English Revolution proved remarkably durable and were influential on a number of Englishmen in these years. In the short-term, however, the radical ideas and movements which offered a different and contrasting future for English society derived from more immediate circumstances.

Throughout these years a host of western societies was troubled by upheaval and change, to the extent that historians readily accept the concept of the 'Age of Revolution' to cover western experience in these years. There is no doubting the ubiquity and the penetration of those upheavals. So far-reaching were they that they even loosened the chains that had for centuries bound the black slaves in parts of the New World.[1] The revolutions which transformed western life – the American and French Revolutions – were, whatever their inspiration, major changes in the *political* structure of those societies. And this is a reminder (if one is needed) that politics form an integral part of the contemporary social experience. In the writing of modern English social history, however, there has developed a tendency to minimize – or ignore – political life. In part, this may be because England did not directly experience *political* revolution. True, the American colonies broke away and transformed themselves into the USA – but that took place 3000 miles away. Closer to home, the French Revolution clearly influenced the development of English politics – but not through violent upheaval and change. The greatest force for change in England was the complex transformation we describe as the 'industrial revolution'. It has, therefore, seemed reasonable to many social historians to write of England in these years as if politics were mere 'noises off stage', while the major transformation of economic and social life occupied centre stage and therefore ought to command the historian's undivided attention. Not surprisingly then, the most common historical perception of England in these years – of a nation subjected to economic and social upheavals while remaining relatively immune to political change – has, to a large degree, determined the nature of historical reconstruction. Even without following Trevelyan's precept, the writing of modern English social history tended to evolve without politics.[2]

It is perfectly clear that such a perception of English history is not only artificial but actually distorts our image of the past. There were no clear divides between political and social experience; one was shaped, in large part, by the other in a mutually and reinforcing interdependence. It is, for instance, hard to understand political changes in these years without coming to terms with the social difficulties taxing the political system. Those men involved in politics – high or low – were wrestling with problems created by social and economic change.

As we have seen, the most complex of contemporary problems were posed by population growth and urban change. But the political difficulties faced by national and local government throughout these years also derived from a growing radical critique of the existing political system – a critique which was itself partly determined by contemporary social problems. Indeed, it would be possible to write a history of these years from the viewpoint of the radical movement. Between 1776 and 1848, there developed the first effective radical (later working class) movement in modern English history; its purpose to transform the structure of politics in order to provide a voice for, and defend the interests of, that great majority of the people who remained beyond the political pale. The poor, the dispossessed, the struggling artisans, the early proletariat – and even the unenfranchised middle class – came to be attracted to aspects of the radical movement because it promised them not merely a political role, but, more important perhaps, it offered a possible solution to their encircling social difficulties.[3]

A new radical ideology seeking solutions far wider than the mere tinkering with representation to Westminster blossomed after 1815, but its origins can be traced to the mid eighteenth century. As early as the 1760s, there had developed, in and around the support for John Wilkes, a far-reaching critique of the system of representation and government. Among men of substance and attainment – who often lacked a corresponding political power – there emerged, in the counties and in London, political organizations which demanded reform of representation and more rational and less corrupt conduct of politics and economics. These arguments were strengthened and extended by the events in North America after 1776. Those American debates about representation and taxation, about the relationship between colonies and metropolis, were similar in kind and tone to many of the arguments within the Wilkes movement. They were also attractive to similar social groups; the upwardly mobile men of business and enterprise, the skilled and educated artisans, and those members of the urban 'mobs' who could be persuaded of their own interest in ending what, on both sides of the Atlantic, seemed to be a corrupt denial of hallowed British constitutionalism. It was no accident that Wilkes and his backers were sympathetic to the American cause (and vice versa). Furthermore, the disastrous and costly British conduct of the war in the American colonies (1776–83) helped confirm its critics' belief in the pressing need for political reform. As if further stimulae were needed, the swirl of reforming debate in the newly-independent USA sent eddies of intellectual reformism back across the Atlantic. By the late 1780s, Britain had witnessed her valuable mainland colonies transformed into a democratic republic, buttressed by a structure of local and national democracy (the existence of slavery notwithstanding) which went far beyond the ken and aspirations of British reformers. To Englishmen, struggling – in rural or urban communities – with the political burdens of governing a turbulent, expansive and migratory people, and for whom the difficulties of guaranteeing adequate food, wages – or basic stability – were visibly mounting, the American solution provided food for political thought.

More significant still, perhaps, was the impact of reforming ideas *beyond* the pale of formal politics. Even before the seismic impact of 1789, many thousands of ordinary working people had begun to find their political appetites whetted by a heady mixture of economic dislocation and those elementary political truths washing eastwards from America. The first spasm of modern mass politics had begun to reveal itself in the 1780s, notably in the massive and unprecendented support for the campaign against the slave trade; a campaign that used the language and tactics of popular democracy and that called into being a new and relatively untried constituency to persuade Parliament of the need for political change. But even this was upstaged by the consequences of the upheavals in France after 1789.

The English response to 1789 was hesitant and mixed, partly because it remained unclear what would be the political outcome. By 1791, however, the domestic political debate stimulated in large part by Edmund

Burke's *Reflections*, the spreading chaos in France and the impact of serious economic difficulties in Britain, began to arouse political awareness among distressed artisans and workers in London and other industrial centres. Among such men there rapidly developed and flourished the reforming corresponding societies, whose beliefs were founded not directly on the principles of 1789, but upon the English text, Tom Paine's *Rights of Man*. Paine's publications sold in unprecedented numbers and formed in effect the origins of that tradition of cheap, popular and radical literature that was to characterize urban politics throughout our period and beyond.[4] In its early years, the *Rights of Man* became the 'bible' for those massive popular societies – and the even larger crowds attracted to public meetings – which dominated urban politics throughout the 1790s.

The popular radicalism of that decade was unique in many ways. It had unusually widespread support, was national in its appeal, articulate and reasoned in its demands, and urban and plebeian in organization. Indeed, these were the very factors that so alarmed William Pitt's beleaguered government which found itself confronted by the massive urban support for political change among people customarily dismissed by those in authority as devoid of political rights (and incapable and therefore unworthy of arguing for them). It was easy – and politically persuasive – to dismiss English radicals of the 1790s as mere 'Jacobins'; the treasonable agents of French revolutionary ideals, anxious to plant the seeds of disruption in the volatile urban areas of Britain. It was also untrue. None the less, the allegation had the advantage of persuading Parliament to pass the repressive measures needed to outlaw and block the development of open radical politics. Furthermore, this repression was applied with equal vigour to organized industrial agitation. Thus, as the nineteenth-century dawned, organized reform and industrial agitation succumbed to repressive legislation, though there survived elements of an underground tradition which sought to preserve and extend the demands for political and industrial change.

Such repression – backed by punitive laws which threatened imprisonment and transportation – was accepted by many because the nation was at war (after 1793). Yet, it was that very war which compounded the nation's social and economic woes. The hunger of 1795, the naval mutinies of 1797, the upheavals in the textile industries at the end of the decade, and, perhaps worst of all, the bloody Irish rebellion of 1798 all (and more) generated waves of radical political activity which convinced the governing orders of the need for unyielding and hostile responses.

Thus, the nation entered the new century in a divided state: plebeian organizations were under the ban, while governments of varying political permutations viewed restrictions on political liberties as the key to securing domestic harmony at a time of unparalleled wartime stress and upheaval.

The pattern of radical politics familiar in the 1790s re-emerged, with the added vigour supplied by worsened urban circumstances, in the miserable years that followed the peace of 1815. The third of a million men discharged from military service, many after years of privations, were rewarded (like so many in 1918) by unparalleled domestic and economic troubles. The war, it must have seemed, had been the harbinger not of better things to come but of heightened social tension and misery. This was particularly a problem in the new industrial towns, many of them troubled by immigration from rural England and Ireland, taxed too by the upheavals of industrial change and post-war readjustment. And it was these same years which were thoroughly infiltrated by the popular radical traditions based on the printed word, and rooted in democratic and industrial organizations. Peace saw a recrudescence of radical agitations – on an even larger scale. From 1815 to the mid 1840s, there were to be periodic spasms of urban unrest, of varying intensity and in different locations, which threatened the very fabric of political stability. It must, at times, have seemed to contemporaries that the nation was stumbling aimlessly from one crisis to another. And yet, by the end of our period, England had entered on a period of relative tranquillity which, even in contemporary eyes, was seen as a major breakthrough towards a more stable (though hardly more equitable) society.

As peace settled in 1815, in the words of Samuel Bamford, a Middleton weaver, 'the elements of convulsion were at work among the masses of our labouring population'.[5] One major point of contention was the Corn Bill of 1815, seen by urban working people as a tax on their food in order to safeguard the well-being of the landed interest. There was, it is true, a notable identity of interests between middle- and working-class radicals, but *the* most striking political feature of these post-war years was the emergence of a new kind of plebeian, urban politics. For all the violent eddies of food and anti-machine riots (so characteristic of former times) plebeian politics had entered a new phase, its mature style and ambitions shaped by the urban environment. The kind of politics that had once been distinctive of (though not entirely unique to) London, now began to emerge from many of the newer urban and industrial areas. This was particularly so with the case of political education, fed by a growing number of self-consciously didactic popular newspapers and publications

– most notably William Cobbett's *Political Register*. Again, to quote Bamford:

> At this time the writings of William Cobbett suddenly became of great authority; they were read on nearly every cottage hearth in the manufacturing districts of South Lancashire, in those of Leicester, Derby, and Nottingham; also in many of the Scottish manufacturing towns.

Bamford and others believed that such popular journalism was responsible for channelling plebeian protest *away* from the old turbulent stream: 'the labourers read them, and henceforth became deliberate and systematic in their proceedings'.[6] In fact, the development of this rational plebeian literacy had, as early as 1791, alarmed men in government; an alarm which inspired the stamp tax on newspapers, which, in its turn, became the object of radical hatred until its repeal in 1855. Between 1830 and 1836, more than 700 people were prosecuted for selling unstamped newspapers. These journals were, as government and radical alike both knew, the very lifeblood of plebeian radicalism and their importance was illustrated by the central role of the *Northern Star* in the Chartist campaign of the 1830s and 1840s.

The immediate post-war discontents – with urban violence, threats of conspiracy (notably in London), the proliferation of radical clubs, protest marches and massive public meetings – were sustained by the plight of working people throughout the Midlands and north. But the story was also perniciously distorted and misled by government spies. These years culminated in the 'Peterloo' massacre of 16 August 1819 in Manchester, when, in an attack on a peaceful crowd demanding reform, the yeomanry killed eleven and wounded 600. The event horrified impartial opinion, but confirmed the plebeian radical mood that working people could trust only their own kind, a feeling easily fostered by the sense of physical isolation already established in many industrial areas of the north. The process of urban and industrial development which isolated whole sections of working-class life, the 'zoning' of different social classes in different parts of the towns, was now paralleled by a concerted political threat to isolate and neutralize plebeian politics, a process culminating in the Six Acts of 1819–20 which, like their forebears in the 1790s, rendered illegal the main features of popular radical politics.

After a fashion, the Six Acts were successful and heralded a new decade of relative social calm. Radical politics continued, of course, but in less strident a form. Popular radical journalism survived, notably in the bigger industrial cities, while many radicals furthered their cause through varied educational activities in the urban and industrial areas. Sectarian organizations flourished, with their own particular emphasis on self-education and self-improvement. There were artisan-based Zetetic mutual improvement societies throughout the urban north; heirs to Paineite Jacobinism, they belonged to an older pre-industrial tradition that was ill-suited to the rapid proletarianization of the industrial regions. None the less, from this tradition, the English radical cause developed its distinctive ideology (later still to become socialist).[7]

At the level of structure, however, what came to characterize working-class life were industrial and trade organizations. Paramount were the combinations and unions (legalized in 1825) which began to defend the sectional interests of growing numbers of working people, especially in industrial communities. It was from this union base that there emerged in Lancastrian and Yorkshire towns a powerful working-class urban presence to secure legislative protection for working people. Consequently, by the mid 1820s, we can see a fusion of radical and trade organizations and ideas to promote the political and industrial interests of working people throughout the country – but with its base and greatest strength in the industrial towns of the north. And, as a complement to those people wishing to transform society – to make it more equitable in political and industrial terms – there emerged a new breed of men, led by Robert Owen, who sought to throw over the imperfections of existing industrial and urban life by creating new, ideal communities. Such new Jerusalems enjoyed mixed fortunes, but they provide a clear indication of what their founders rejected, namely, the increasingly harsh industrial environment that seemed resistant to humane correction and almost beyond the ameliorative power of men.

Whatever the problems endemic to the new (and old) urban areas, towns were the ideal breeding grounds for the development and conduct of mass politics. After all, the popular radical traditions in London, throughout the seventeenth and eighteenth centuries, had provided a political forum for untold legions of labouring men traditionally kept outside the pale of formal (and propertied) politics. Radical politics were encouraged by, and thrived on, the urban and industrial experience; the difficulties of the densely-crowded urban environment periodically created widespread unrest and demands for redress. This is not to claim that all distressed town dwellers were, necessarily, supporters of each and every radical cause. On the contrary, there was throughout these years and beyond a powerful and vociferously loyal and obedient streak in plebeian communities which could, at crucial moments, be rallied to the defence of the political and social status quo by agencies of local or national government. The cry of 'Church and King' could be guaranteed to marshal public support at critical moments.[8] None the less, towns provided the basic human and intellectual ingredients for the chemistry of mass and radical politics.

At the obvious level, it was, for instance, a simple matter to drum-up a crowd and organize private or public meetings. And, as towns grew, it was as easy to summon a vast gathering in Sheffield or Manchester or Birmingham, just as had traditionally been the case in London. The 60,000 gathered at Peterloo is an obvious example. The concentration of population made it possible to contact, inform, marshal and persuade armies of people; converting them to a political, religious or trade argument. In rural life, with its diffused and scattered population, this was obviously much more difficult. In urban life, as a 'face-to-face' society slowly gave way to a more 'atomized' life, where the employer and manager of labour became ever more distant from employees, working people could lead their social and political lives beyond the ken and the influence of their betters. Deference and intimidation began to give way (though not universally or completely) to independence and greater antagonism. And this was facilitated by the development of those political and social institutions of urban life which were themselves a response to urban problems.

This was most overtly the case with radical organizations or trade unions. But the towns and cities made possible a flowering of a host of other institutions which, though non-political in a formal sense, often had major political repercussions. This was the case with the churches, chapels and Sunday schools which, far from having the deadening effect so often alleged, were sometimes responsible for the infusion of a spirit for change among thousands of urban people.

This can be illustrated in particular by the campaign against the slave trade and slavery. The tactics and vocabulary of anti-slavery had been well-rehearsed in the 1780s and 1790s. But, like radicalism, anti-slavery had fallen victim to repressive fear of Jacobinism after 1792. From the mid 1820s, however, it entered a new phase, anxious to recapture public attention, and channel popular feeling towards freeing the three-quarters of a million black slaves in the British Caribbean. Once more, anti-slavery was able to break out of the conventional restraints of formal, propertied politics and to capture vast crowds of popular support. It was able to do so for two prime reasons, both closely related to urban and industrial change. First, anti-slavery used as its base the rapidly proliferating network of churches, more especially the non-conformist chapels, that were making so profound an impact in new working-class communities. And, second, the imagery and language of anti-slavery was of immediate and direct application to the domestic political debate, particularly about industrialization. Indeed, the trade union and Factory Act movement swiftly adopted the language of anti-slavery. Factory workers were portrayed as slaves to a corrupting and immoral system; factory owners were cast in the role of slave drivers, lost to human kindness and sensibility. Furthermore, there was a powerful and growing sense in which slavery was viewed as irrelevant; a survivor from a former age, a now defunct economic system which was out of kilter with the times.[9] Working-class complaints against, for instance, the Corn Laws as a tax on the working-man's diet applied with equal force to sugar, for the sugar duties kept sugar prices artificially high in order to safeguard the economic interests of the slave-holding plantocracy. Thus, Caribbean planters and English landowners seemed, to their political opponents, united in clinging to an economic protective system designed to penalize the poor and the ill-paid English worker.

It was, however, the religious arguments against slavery which became more strident. Those churchmen – notably Methodists and Baptists – most active among the slaves were, at the same time, attracting ever more working people into their urban congregations. And news from dissenting ministers regularly brought news from the West Indies directly into English working-class communities, many of whose members at the same time felt themselves to be the victims of a wicked industrial process seductively similar to slavery. It was, then, understandable that working men, women – and children – were won over to anti-slavery in unprecedented numbers. By the late 1820s and early 1830s, the only restraints on the size of anti-slavery meetings was the capacity of the meeting-place. People signed anti-slavery petitions in unimagined numbers; more, in fact, than for any comparable radical issue – including Chartism.

Public meeting-places were essential for anti-slavery, and increasing urban facilities meant that they were always available. Churches, chapels, town halls, Leeds' 'Coloured Cloth Hall', assembly rooms, taverns, courts – even a music hall – were all made available to the anti-slavery cause. Paralleling these public meetings there was an overwhelming outpouring of anti-slavery literature; of tracts, newspapers, magazines and handbills, spilling from the presses to satisfy the growing literate appetite of urban society. However distant and remote the slaves, they none the less became, by the late 1820s, the object of an amazing upsurge of public feeling, among all classes, and both sexes, throughout Britain.[10]

The extent of anti-slavery feeling was closely associated with the transformation of urban society; notably the changes in organized religion, the spread of popular literacy, the development of industrial work and the diffusion of reforming sensibilities which were encouraged largely by the wider reforming agitation. By 1830, it was generally accepted that black slaves ought to enjoy the 'rights of man'. Yet, a mere thirty years earlier, Englishmen had been transported to Australia for claiming such rights for themselves. The vocabulary of the democratic revolution of the late eighteenth century had firmly established itself as an unexceptional principle in English urban life by the time of Victoria's accession.

The development of a commitment to democracy in urban society was of major significance for the wider progress of reform. Indeed, the wave of reforming sensibility that swept across England from the late 1820s, culminating in the reform of Parliament in 1832, was carried along on a number of issues (which often became indistinguishable one from another). The heady climate this generated was heightened, for good and ill, by the threat (and in some cases the reality) of disorder and upheaval, and then, as if to cap it all, by the terrorizing attacks of the cholera in 1832 itself.[11]

The crisis years of 1830–2 revealed the manifold problems posed by urban society, but in a context where the worst disturbance of all – the Swing Riots – were primarily rural. It was, however, the formal popular radicalism, nurtured in the 1820s, which emerged into a potent political force in 1831–2. Massive crowds, a ferment of local radical groups in Midland and northern towns, compounded the plebeian agitation in the capital. What we see in 1830–2 'is an organised, independent working class presence, sometimes working in co-operation with radicals from other classes, sometimes standing alone in opposition to all'.[12] To an alarmed propertied system this was dangerous enough. As if to confirm these same men's fears of plebeian politics, in places radical politics spilled out into violent public disturbances. In Derby and Nottingham, but especially in Bristol and Merthyr Tydfil, riots erupted on a scale that had a chilling resemblance to the Gordon Riots fifty years earlier – but this time in provincial cities. If urban unrest on the eighteenth-century scale was to persist, the problem was, quite obviously, much more serious now that the number and size of towns accentuated the basic problems.[13]

There was in 1831–2 a volatile political atmosphere. The urban violence coincided with – and fed into – renewed agitation for a free working-class press, the revival of trade unionism, and a growing insistence in the industrial north on protective factory legislation. When the campaign for the reform reached its height, a Parliamentary committee took evidence about the worst abuses of factory labour, and from this subsequent data there developed a growing awareness of the indignities and cruelties endemic to industrial progress. Symbolically, the expression

'Yorkshire Slavery' was coined to focus attention and revulsion on barbarities at home. Moreover, politics were becoming divisive. Despite the support of this or that local Tory or Whig, more and more working men throughout the country were coming to the view that their best interests were served by themselves, and not to expect help from either party. From this self-reliance there was to blossom that working-class political independence which was sharpened by the rebuff of the Reform Act of 1832 (which made no political concessions to working men), swiftly followed by the insult of the 1834 Poor Law, and then the slide into economic difficulties in the late 1830s.

When the poor law was introduced into the north in 1837, the violent response it provoked underlined the uncertainty and volatility of governing the urban and industrial areas. The old 'short-time' committees were instantly transformed into anti-poor law committees able to muster large meetings, to issue fierce denunciations, and, at times, to stimulate physical violence. This opposition undoubtedly blunted and delayed the implementation and operation of the new act, and anti-poor law feeling, itself heightened by the economic gloom that settled on industrial regions, led directly to the formation of Chartism. Here, yet again, was another articulate, reasoned, mass movement aimed at Parliament's doors, and finding its succour and strength primarily in urban life. It was, however, the disillusionment with past experiences that infused Chartism with its acerbic quality. Chartism fed on the reality of the 'Great Betrayal' (of 1832) and inherited the sharpening sense of class-consciousness and the bitterness of the anti-poor law movement. From this unstable mix, there was to develop the first major mass working-class movement in modern English history.

Many of the demands and tactics of the Chartists seemed a reprise of radical politics of the late eighteenth century. But Chartism went beyond that, for it was permeated with a bitterness derived from the intervening years and was fuelled by the distinctive problems of the varied urban communities which spawned Chartist activity. Public meetings attracted vast crowds from the bigger cities while the movement readily acquired 1,280,000 names for its first major petition. There was talk of an uprising and, in places, serious violence did occur. But the first wave was rebuffed and silenced by mass arrests in 1839–40. Yet, within two years, as industrial troubles multiplied, Chartism re-constituted itself, mustering 3,317,752 names to a new petition for reform. Its rejection coincided with major industrial depression throughout the northern industrial towns in 1842. Once more, however, Chartism was rebuffed, only to revive for a final assault on a resistant Parliament in 1848. Then it flourished in the shadow of the revolutions which plagued the major cities of western Europe, a fact that compounded the apprehensions of the government. In 1848, Chartism seemed, to its opponents at least, to be part of a wider European phenomenon threatening to plunge all the capital cities into a crisis of general disorder. Once more, however, the peaceable Chartist movement retreated in the face of Parliamentary obduracy – and melted away.

Despite its undoubted failures, Chartism also left behind a solid contribution. Throughout major towns and smaller villages, it left a legacy of radical agitation and self-education, most notably organized around the literate habits and debates, which were themselves fostered by the Chartist *Northern Star*. Moreover, the heightened political awareness of the Chartist years survived among many local men and women, to be redirected into a host of subsequent political and social campaigns. While being rootedly and unshakably working class, Chartism did not comprehensively embrace the totality of plebeian social and political experience. It often competed with Owenism and, most obviously, with trade unionism. And there was another countervailing force – admittedly not as strong – seeking to unite the middle and working classes, but it was only at the end of our period that the first effective political bonds between those two social classes were being struck, in the National Parliamentary and Financial Association. In the words of Edward Royle, 'the first, tentative steps had been taken on a new stage of the history of radicalism and reform which, in the late 1850s were to lead towards popular Gladstonian Liberalism'.[14]

The years 1815–48 had witnessed a series of profound political uncertainties and upheavals; waves of popular industrial and political agitation attendant upon major economic dislocations. At times, it seemed unlikely that the political system would survive popular radical pressures. And yet, by mid century, English urban society seemed to contemporaries to be more stable, peaceable, and more governable than at any point in the living memory. The decline of formal radical agitation is itself a symptom of that change and, in large part, can be seen as a result of the maturing of an urban and industrial society.

To argue that the mid century industrial economy had begun to yield more of its benefits to ever more people is to offer far too crude a view of what was clearly a complex phenomenon. The radical movement since 1776 had not, after all, been merely a matter of hunger or wages, but had embraced an alternative and fairer political view of the world expressed in articulate assertions of political rights and obligations. While it is clear that these ideals long pre-dated the modern urban world, the complex ramifications of urbanization gave them added edge and urgency. Similarly, many of the seminal radical movements and leaders, from the corresponding societies of the 1790s through to Chartism, belonged to (and were spawned by) the old trades and skills, many of which found themselves increasingly threatened by urban and industrial change. But these organizations and ideals were now transmuted into the stuff and structure of modern urban mass politics. The radical movement in these years may seem to have striven in vain, but it was undoubtedly crucial in creating, within the differing context of English towns and cities, the ideas, organization and experiences that survived into the second half of the century and formed the basis for modern mass politics.

Men in office may have felt by mid century that they – and all they represented – had survived remarkably well, with only minimal concessions to popular demands. Yet, their survival cannot be explained purely in terms of their own tactical successes against radical pressures. It is true that the English government was successful in maintaining its grip over what seemed at times to be an increasingly volatile urban population. But, to do this, it had to reform not only the nature of central and local government, but also come to terms with urban growth by establishing completely new policing systems. And yet, when the English people seemed more stable by 1851, this was not uniquely a function of the physical or policing dominance of their governors. The English had, by the mid nineteenth century, become an altogether more governable people and, in large measure, that was because they had become a nation of town dwellers, ever more accustomed to the disciplines and codes of life in the urban environment. Just as the industrial labouring force had come to terms with the new labour disciplines and requirements of work dictated by the machine, so too had the wider urban population accepted the different restraints of town life. It may seem paradoxical, but no sooner had the final Chartist challenge faded away than contemporary Englishmen began to remark on the tranquillity of the English people.

If, in the short term, the two generations of radicals, spanning the years between the French Revolution and Chartism, had been unsuccessful in creating the more equitable and just system that lay at the heart of their ideology, they were none the less influential in establishing a structure of mass politics ideally suited to town life and which was to survive in its basic form into the twentieth century.

8 *The people subdued: law and order in an urban setting*

At the beginning of our period, and for a long time before, the English were infamous as a turbulent and ungovernable people.[1] By the mid nineteenth century, however, this image had changed quite fundamentally – though very recently. At the end of our study, the state had called into being the apparatus of policing and imprisonment familiar today. But this alone does not completely explain the transformation brought about in the behaviour of the English town dweller. Moreover, for much of our period the English were notoriously volatile, prone to frequent, widespread and severe bouts of collective violence over a wide range of social and political issues. Riots, mob action, popular collective violence in pursuit of a political or economic goal constituted a form of political activity (in some senses, the only one available) which local and national governments had long been accustomed to dealing with, however unhappily. English towns and cities were especially susceptible to the destabilizing impact of collective violence. And, as towns multiplied and expanded, the problems of how to control urban people, especially in times of hardship, were multiplied throughout the country.

The most extreme problem was posed by the urban mob, a feature of English life throughout these years. In the story of organized radicalism for example, from Wilkes to Chartism, each chapter was punctuated by mob activity. At times the mob came to symbolize a wider political issue. Indeed, in some instances, it is the mob which is the best-remembered feature of a particular movement. In the 1760s and 1770s, for instance, the Wilkes' phenomenon is most graphically remembered through the London mob, threatening violence and destruction on the heads and houses of its opponents. Abrogating to themselves the symbols, styles and vocabulary of legitimate authority, the Wilkes mobs were, in effect, the Lords of Misrule, turning the political world upside down and imposing, through their collective strength and the threat of disorder, a political order of their own making.[2] While the Wilkes' affair may have been extreme, it was a striking manifestation of the wider and inescapable fact that major and frightening turbulence was a feature of English urban political life throughout the eighteenth century. This was no longer the case by 1851.

It is, naturally enough, difficult to disentangle strictly political issues from, say, religious or economic ones. In the making of collective violence, none the less, formal politics – at elections and the hustings – were invariably accompanied by violence, certainly throughout much of the first half of the eighteenth century.[3] This was particularly so in London which, because of its size, posed severe problems of urban control. The population in general, and the crowds it spawned on specific occasions, were highly volatile, unpredictable, and, ultimately perhaps, uncontrollable by friend or foe, a fact more than amply illustrated by the horrifying violence of the Gordon Riots of 1780. Ostensibly anti-Catholic and pro-Protestant, and beginning with the well-tried tactics of extra-parliamentary pressure, the vast crowds assembled in the humid summer weather. Sections peeled away to attack Catholic property, and there followed days of escalating violence and destruction in which 285 were killed or died later. Damage to property was enormous, and, of course, the law eventually took its own toll of human life.[4]

However rational – and selective – the Gordon rioters' political aims, they cast a shadow over subsequent extra-parliamentary pressure groups and reinforced, among men of property, an acute awareness of the fragility and precariousness of their political hold over London. The machinery of urban control was, quite simply, inadequate to the task, and while this had been generally true of the capital for centuries, the problem was accentuated by the massive rise in population and the intensification of human problems in the capital. Furthermore, this feeling of self-doubt was quickly replicated throughout the expanding urban areas in the late eighteenth century where local men of authority found the mechanics of local government (i.e. the forces of law and order at their disposal) quite inadequate to the varied tasks of governing and controlling. In the case of Manchester, for instance, where, throughout the 1780s and 1790s, the expansive textile trades constantly attracted an influx of migrant people, the town was governed by an utterly inadequate Medieval Court Leet. Symptomatic of its incapacity to govern effectively was the rapid proliferation, in the 1780s, of private associations to carry out the essential functions of law and order.[5]

Whatever the weaknesses or shortcomings of the structure of urban government, they were thrown into sharp relief by the Gordon Riots of

1780. While it may be thought to have been a uniquely metropolitan phenomenon, these riots cast a long shadow across the governance of urban England for decades to come. Indeed, memories of 1780, and the ease with which large-scale political gatherings could be transmuted into uncontrollable mass violence, influenced the propertied and governing responses to subsequent generations of extra-parliamentary pressure groups of various kinds. Furthermore, it was a memory which was sharpened by the propertied interpretation of events in France after 1789. By the early 1790s, it seemed, both at home and abroad, that the politics of protest was suffused with an endemic threat to the very fabric of order and stability. However malignant or even vicious the response of Parliament and government to extra-parliamentary pressure (notably plebeian and radical), it needs to be remembered that the memories of 1780 and 1789 left incurable scars on the political psyche of a whole generation. Many thousands of propertied Englishmen concurred with the sentiment of Joseph Brasbridge, 'From that moment . . . I shut my ears against the voice of popular clamour.'[6]

Political violence persisted, stimulated in the war years by rising distress and hunger – the very forces which encouraged the rise of popular radicalism. The 1790s were punctuated by violent incidents and threats (notwithstanding a large standing army), ominously spilling out into the Midlands and northern cities. It even reached into that bastion of British security, the Royal Navy, in the mutinies of 1797. But it was, yet again, London which seemed most vulnerable – its population of $\frac{3}{4}$ million stretching to the limit inadequate government and policing systems. Under the shadow of events in France and faced by the emergence of an unprecedented radical movement, the British government became obsessed with social stability in the capital. Indeed, it has been argued 'that popular disturbances in the capital remained the first concern of any government in the late 18th century and early 19th century England'.[7] It is true that there was nothing remotely comparable to the Gordon Riots, but the 1790s witnessed massive public demonstrations which were actually larger than the Gordon gatherings. While the reality of violence did not materialize, the threat of it continued to nag and worry the governing classes. Indeed, the radicals had regularly to deny accusations of violent intentions:

> One of the fundamental principles of this society [said the LCS] and a lesson we have industriously inculcated is that riot, tumult and violence are not the fit means of obtaining redress of grievances.[8]

Their propertied opponents chose not to believe this disclaimer, a view apparently confirmed by the rise of Irish revolutionaries in England, the naval mutinies, Irish rebellion in 1798, and the periodic acts of violence which seemed attendant on (though in fact clearly not caused by) formal urban radicalism. The hearts of a whole generation of governing propertied Englishmen had been incurably hardened by the Gordon Riots which became, in effect, an ill-fitting and inappropriate gauge against which subsequent reforming and mass movements were judged – and normally condemned.

Throughout the post-war years, the threat of political violence was rarely far from the minds of national and local government. Men in authority found it hard to shake off fears of public upheaval. Of course, the disturbed social conditions after 1815 (especially in the years 1815–19, 1830–2, 1837–9, 1842 and 1848) seemed merely to add grist to their mill. Now, however, and unlike the years before 1789, this threat was to be found not merely in London, but scattered across the face of urban England. The fears of the governing orders, especially in the localities, were to be measured by the inadequacy of their particular governing systems. Not until municipal reform and the establishment of effective policing systems in the last twenty years of our study, were these fears to be diminished.

It would be wrong to imagine that public disturbances in pursuit of a political aim were the monopoly of radical groups. Ironically, some of the worst urban violence was inspired by conservative and loyalist organizations. This, after all, has been the case with the Gordon Riots. But it reached a peak in the years of the French Revolution, when, worried by France, men of property organized and orchestrated (though sometimes losing control of) a spate of 'Church and King' riots on behalf of the political status quo. The most serious incident was the Birmingham Priestley Riot on 19 July 1791 where violent attacks were made on the property – and persons – of local reforming dissenters (though sometimes to the cry of 'No Popery'). A year later, after the publication of the *Rights of Man*, Tom Paine became the symbolic demon in a revival of loyalist rioting much of which, directly organized by local groups of the 'loyal' propertied, received indirect support from Pitt's government. Time and again, mobs were conjured forth from the urban recesses, plied with ale, stirred in with verbal abuse, and physical assaults launched at the heads and homes of local radicals and reformers. And at the centre of these turmoils of loyal violence there were the symbolic burnings of effigies of Paine. The objects of such physical intimidation varied from place to place – a publisher here, an activist there, and a sympathetic landlord elsewhere – but the common and recurring theme was the vilification and burning of Tom Paine's effigy. Within a 20 mile radius of Manchester, at least twenty-six such Paineite burnings took place in the winter of 1792–3.[9] Such burnings were often turbulent, threatening, and often drunken crowd activities; parades of men past the homes of prominent local reformers, leaving them in no doubt that their ideas were opposed by more physical sections of the neighbourhood. Often, local reformers were physically attacked. Furthermore, we need to recall that the Association movement, responsible for the wave of loyalist violence, was indirectly encouraged and helped by Pitt's government. Anxious to maintain peace and stability, the government and its supporters were not above indulging in organized violence and fisticuffs to intimidate its plebeian opponents. It was – and ministers knew it – a dangerous game to play in times of such distress, and in urban areas troubled by the war.[10]

Although the most extreme forms of 'Church and King' riots had ended by 1794, similar outbursts occurred later when, in the public euphoria of military victory, rejoicing mobs could turn their fickle energy against people who failed to share their enthusiasm. Homes not 'illuminated' to celebrate a victory were often attacked – so too were homes of well-known local radicals or opponents of the war. But, in most cases, the violence was specific, limited and targeted against people who were *known* to be unsympathetic to the mob's rejoicing.

It is naturally impossible to draw up a balance sheet of urban violence; to try to compute the greater popularity of one side or the other. Were the radical issues more 'popular' than the loyalists', for instance? It seems obvious and likely that they did not appeal to the same people. And yet, we need to balance our image of political violence by recognizing that it was used, manipulated – or sometimes erupted spontaneously – by diametrically opposing philosophies. While men of property feared the violence of the mob, they were not averse, at crucial junctures, to manipulating mob violence and directing it against their opponents. Furthermore, it would be wrong to ignore the large sections of urban life ready and willing, for whatever reason, to offer their vocal, physical, and collective strength in the service of intimidating or hurting local reformers or radicals. Urban violence was *not* the monopoly of lawless banditti or of reforming groups anxious to muster public strength to their side. Their opponents were, in many respects, even more adept and accustomed at using the urban mob. Men of substance had, after 1780, come to appreciate the precariousness of such an exercise. None the less, so worrying was the rise of radicalism, that local and national government was prepared to use the mob as an instrument to intimidate their urban opponents.

We have already seen the consequences of the wars for the development of post-war mass urban politics. Understandably enough, there were few aspects of English life unaffected by that generation of international conflict. The wars had drained Britain of manpower and money; no less than £1500 million had been raised in taxes and loans, a burden of National Debt which was as economically burdensome as it was politically embarrassing. While many basked in the memories and mythology of British military prowess, symbolized by Waterloo, many more struggled to survive as those industries, artificially expanded by wartime production, rapidly shed their manpower. And the dislocation was paralleled by an agricultural slump. A deep and abiding frustration and cynicism entered the public mind.

You say that Bonyparty he's been the spoil of all
And that we have got reason to pray for his downfall
Well, Bonyparty's dead and gone, and it is plainly shown
That we have bigger tyrants in Boneys of our own.[11]

But, as we have seen, the attempts at redress were thwarted. The English governing orders proved more resistant than Napoleon's armies.

Not surprisingly then, in England after 1815 there was turbulence in abundance. But the great bulk of it did not stem, as its opponents commonly claimed, from formal radical politics. The traditional violent responses to distress continued – riots, attacks on employers and hoarders of foodstuffs, 'popular taxation' and the wider 'political economy' of the crowd. Food riots, for instance, were more common than radically-inspired violence; mass gatherings of the unemployed more unsettling than reform agitation. Of course, these issues are not so easily separated one from another. What alarmed that government was that radical agitation in distressed communities might be the spark to ignite a major upheaval. Thus, the men governing in the localities regularly took punitive action against radical meetings, unemployed marches, and, of course, against whatever evidence existed of insurrection. Spa Fields, the Blanketeers, Pentrich, and, ultimately, Peterloo, the succession of incidents is well known. Throughout, the poisonous words of government informers served merely to confirm government suspicions; transmuting the legitimate and not unreasonable demands of working people – for work, for food at prices they could afford, and for some measure of political reform – into the symptoms of threatening and subversive insurrection.

After the violence of Peterloo, the government armed itself with an array of repressive measures – the Six Acts – similar to those of the 1790s. Thus, the government entered the rapidly changing world of the 1820s equipped with the repressive legislation of the eighteenth century and haunted by the mentality of an older governing order. Time and again, men of influence and property referred to Peterloo in terms of the Gordon Riots. The governors of England viewed the problems of the early nineteenth-century urban life through eighteenth-century eyes. Significantly, however, their fears had shifted from the capital to the centre of the textile trades, 200 miles to the north.

Given the obstinacy of successive post-war governments, it was perhaps inevitable that demands for reform would create turbulence among the large crowds which were naturally attracted to the seductive calls for change. Leaders of reform agitation consciously used those crowds they could muster to threaten the brittle façade of government resistance. Thus, the reform crisis of 1830–2 took place against a background of threatening – and, in many cases, real – riot and popular menace. Riots, attacks on unpopular individuals, and massive crowds formed a threatening urban context to each and every stage of the unfolding political drama. A crowd of 100,000 gathered in Birmingham to protest against the Lords' rejection of Grey's Second Reform Bill. A mob in Derby attacked the local gaols – the Riot Act was read – and disturbances spread throughout the towns of the east Midlands. But it was Bristol which suffered the worst of the reform riots, with massive looting and burning and large-scale deaths and injuries. Troops specially called into the town lost control of the city to the mob. Local order simply collapsed in a confusion of indecision, and the uncertainty of where responsibility lay, between magistrates and troops. It was a truly eighteenth-century problem, revealing the inadequacy of an outdated structure of local government when faced by massive urban unrest. London now had its own professional police force; the Bristol riots confirmed the urgency of establishing a modern policing system throughout urban England.[12]

The climax of the reform agitation, in 1832, was much less violent than the year before. But threat of the urban mob was an ever present factor; a card (albeit a joker) thought to be in the radicals' hands – a persistent nightmare for government. Moreover, it was a nightmare accentuated by another, even more deathly, reminder of the imperfections of urban government. If the riots confirmed the need for effective policing, the first wave of the spasmodic cholera illustrated an even more alarming problem of urban life, the ease with which disease could sweep through the squalid urban environment. It may be true that there were few problems of urban life in the nineteenth century not experienced in London in the seventeenth and eighteenth centuries. Now, however, they were scattered across the land. Characteristically, the cholera first arrived on its deathly march westwards in the new town of Sunderland in September 1831; Gateshead in December, and then on to London in February

1832. By November 1832, it had fanned throughout Britain, appearing eventually in 431 towns and villages and killing 31,376 people. It exacerbated social tensions, spread panic and alarm throughout urban life, and added its own tormented anxiety to the reform crisis.

The disease was a particularly horrible one; death was cruelly painful, and the disposal of the corpse a pestilential nightmare. Whatever ingredient cholera added to the urban political crisis of 1831–2, it also pointed to even more fundamental flaws in English urban society. Urban life seemed to provide an ideal habitat for this and many other catastrophic diseases, and medicine seemed incapable of solving urban life's fundamental medical problems. Once again, popular anger and fear took the form of riot, against doctors or local hospitals; becoming yet another reminder, if one were needed, of the survival of old English habits – of the resort to collective violence by urban people at times of tension or distress.[13]

Violence also continued to be a political response throughout the 1830s and 1840s. At times, however rarely, it posed a major challenge to local authority and the forces they could summon via the Home Office. The most obvious and influential of violent outbursts in these years was not, however, urban at all, but the 'Captain Swing' riots which troubled the agricultural south between 1830 and 1832. There had been, it is true, periodic outbursts of rural unrest in the years since the war, but it was the peculiar combination of poor harvests, heavy rural unemployment, miserable prospects, in conjunction with the heady reforming crisis of 1830 which proved so volatile a mix.[14] Machinery-wrecking, arson, rick-burning, assault and robbery all spread through the southern agricultural counties, the contagion of violence fanned by bands of travelling men. The national government was caught flat-footed by its preoccupation with the troubles in the industrial regions. By December 1830, however, government had reasserted its control, and something like 2000 people had been arrested. Eventually, 252 were condemned to death (only nineteen were executed), 500 transported, and 600 jailed. It was as if the system of control was savage in its reprisals in direct proportion to the weaknesses of its hold. Rioters could expect – whatever their short-term gains – the full severity of contemporary law and punishment. But this had always been the case, and yet, full knowledge of the dreadful punishments – for individual or collective violence – was as savage as it was ineffective as a deterrent, a fact amply confirmed at a number of critical junctures in the 1830s and 1840s.

In 1831, for instance, there was a major armed insurrection in Merthyr; 300–400 armed men, supported by thousands of local people, took over the town for four days; their losses eventually ran to two dozen. It was, by any standard, much worse, more severe, violent and threatening, than Peterloo, and was only put down by the hurried arrival of 800 troops.[15] In an England accustomed to viewing political events from London, events on the remoter fringes of Britain warranted little public coverage (notwithstanding the acute political concern within government itself).

More alarming because more widespread, was the violence against the 1834 Poor Law; violence which erupted from 1835 onwards, initially in rural parts, in East Anglia, the south and the west country. This, however, was a spasmodic violence, easily contained and ineffective in halting the establishment of the new and deeply hated poor law system. Much more serious was the resistance in the northern towns from 1836. Organized by Tory radicals, the northern anti-poor law movement was able to tap the fears and hostilities generated by the depression in the textile trades. Officials were threatened, massive crowds convened (one in Yorkshire numbered between 100,000 and 250,000), the Huddersfield workhouse was wrecked and massive anti-workhouse crowds were paraded throughout the West Riding in 1837–8. By 1839, however, the threats had subsided, partly in response to concessions made by local poor law officials, but partly as the emergent Chartists lured local men into their politically-comprehensive movement. Indeed, the anti-poor law movement was instrumental in infusing resolve and radical manpower into the Chartist movement.

Chartism was the most serious challenge to the political system since the heady days of the corresponding societies in the 1790s (to which Chartism owed much of its intellectual debt). Although it is too simplistic to divide Chartism into 'moral' and 'physical' force schools, there were undeniable (and self-confessed) groups of physical force men among their ranks. It was possible to defend the rights to physical resistance as a traditional, hard won, and valuable political right, rooted more particularly in the struggle against Stuart despotism in the seventeenth century. In May 1839, *The Chartist* argued that 'Englishmen have a right, in extreme cases, to have recourse to free themselves from an unendurable tyranny. . . .' Violent resistance was, of course, a dangerous game to play, 'a thing not to be lightly had recourse to. . . .'[16] It seemed to many that collective political violence had actually worked, in the case of the anti-poor law campaign in the north. However, not only was the use of violence dangerous and of dubious validity, but it was politically uncertain and possibly counter-productive; an explosive substance which could destroy careless handlers.

None the less, violence characterized certain Chartist episodes. At Newport in South Wales a rising in November 1839 led to the deaths of twenty-four, and the swift transfer of troops into the area – partly by railway.[17] Other uprisings were muted, notably in Bradford and Sheffield, but a wave of arrests, trials, transportations and imprisonment had, by the mid 1840s, provided an effective brake on the Chartist urge to violence.

This impulse to violence alienated those sections of middle-class opinion which, not unfriendly to the arguments and theories for political reform, nevertheless felt their property and security threatened by the unleashing of violence. After Newport, and later after Chartist violence in Bradford in 1848, the forces of property hurriedly sided with the forces of law and order – against the Chartists. In middle-class provincial society, its emergent business prosperity and domestic life-style firmly rooted in the principle and substance of private property, there was no political attraction in the Chartist lurches towards physical force.

The Chartists were carried along – or cast down – by the tide of contemporary and economic circumstances. The government could, however, rely upon the changing physical nature of English life to help control and subdue Chartist, or any other, unrest. The Chartists were confronted, in the later stages, by governments and their local allies able to dispatch new and disciplined police forces across the face of the country, quickly and efficiently by train. There can be no doubting the remarkable boost to governmental power provided by an effective police system and a new transport network. The need for urban police forces was undeniable in the last twenty years of the eighteenth century. Indeed, throughout those years there was a marked proliferation of private organizations among local men of property to secure their interests and to guarantee their safety. In Manchester, for instance, an 'Association of Proprietors' had been formed in 1786 to organize patrols near their business premises throughout the hours of darkness. In February 1788, the local 'Society for the Prosecutions of Felons and the Receivers of Stolen Goods' boasted that in the past year its successful prosecutions included thirteen transportations, seven imprisonments and five whippings – but only three acquittals.[18] Similar organizations sprouted throughout the country from the 1780s, but they quickly disappeared in the 1840s. The timing of their founding is indicative of the propertied concern about what to do about crime and law and order in the aftermath of the Gordon Riots. Similarly, the decline of the Association, coming fast on the heels of the establishment of local police forces, illustrates the growing sense of security among the propertied orders as 'law and order' seemed to be safeguarded by the police.

Throughout much of our period soldiers had been available to supplement the inadequate local police systems in the unreformed towns. But the militia, yeomanry – or, in the war years, the standing army – offered a blunt instrument which could often exacerbate an urban problem. It was, after all, the yeomanry which caused the Peterloo massacre, and troops which lost control in the Bristol riots. From many angles, it had long been apparent – and was becoming ever more pressing – that urban England was badly in need of a reformed policing system. When, in 1829, Peel pushed through his Police Act, it was substantially aided by propertied fear of political disaffection and mob rule. Thereafter, to cope with what many viewed as endemic (or, at best, latent) lawlessness, more especially in the towns, local police forces proliferated, prompted and sanctioned by new Acts of Parliament (in 1829, 1839, 1844 and 1856). The professional policeman – often an ex-soldier – became an unmistakable figure on the urban and industrial landscape. Even as late as 1856, however, we need to recall that only half the nation's counties had paid officers. But those which had were able to impose a new level of social tranquillity on their people.[19]

However ineffective the eighteenth-century constables had been in affecting the local crime rates, the new professional policemen of the second quarter of the nineteenth century made an indisputable impact. Their immediate – and persistent – effect was felt on the streets as a daily, routine, and inescapable presence. The patrolling policeman rapidly – in some cases, instantly – established himself as an important local figure, able to deal with problems, on the spot, in a way which had rarely been the case earlier. His task, above all else, was to maintain law and order, more especially against what informed opinion in the 1830s came to consider as the 'criminal classes' living at the heart of the urban areas. The belief in a criminal class, whose criminal instincts were hereditary and which formed a distinct urban problem, had far-reaching consequences for penal and policing campaigns at many levels. In the short-term, however, the establishment of police forces had a dramatic impact on crime statistics – and thus on how contemporaries interpreted the nature and problem of urban crime. Inevitably, the policeman's zealous determination to combat the problems of public order and crime (notably against property) produced criminal statistics which, in themselves, seemed to indicate an alarming rise in offences. But, as might be predicted, such figures varied amazingly between different towns and regions – reflecting primarily the direction and pertinacity of the local police administration. This was, for instance, clearly the case with arrests and prosecutions for drunkenness, evidence for which varied enormously depending on the policy of senior local officers.[20] Whatever the results in terms of crime statistics, the police forces brought a new kind of stability to traditionally uncertain urban life. In the worst (i.e. the toughest) 'districts of Salford, Merthyr, Liverpool and London, there was a vigorous police battle for control of the streets. . . .' However misguided or ill-intentioned much of that battle, it was won by the police. Time and again, senior officers were able to document their successes in almost every riot after the Reform Act.[21] There were, it is true, incidents late in the nineteenth century which occasionally rendered even the professional police forces helpless. Compared, however, to the successes, the stability, the orderliness which was gradually imposed on the urban nation, such spasms were henceforth untypical. The police alone were not uniquely responsible for this peacefulness, but, none the less, the English had, by the end of our period, become a governable and orderly people – a remarkable contrast to their eighteenth-century forebears.

One of the most telling effects of the police presence in the second quarter of the nineteenth century was the determined drive against many of the popular customs and recreations that were troublesome to propertied society. This is a theme studied in Chapter 9 but, at this juncture, it is worth recalling that one aspect of 'lawlessness' so frequently denounced was the many noisy, turbulent, and boisterous recreational customs inherited from earlier generations. It was, henceforth, possible to direct the police against such nuisances. In the 1780s, propertied opinion moaned about even the simplest of youthful games. 'Pitch and Toss', according to the *Manchester Mercury* in 1788, 'is perhaps of all other most pernicious to the mind by rendering it unfit for the steady employments of life. . .'.[22] From the 1830s, police throughout England's cities could simply stamp out such troubles.

Understandably, many of the street games that were so widely disliked, and consequently punished by the new police, were children's or young adults' recreations. It was quite obvious that large numbers of children would find themselves caught in the new police nets from the 1830s, merely for doing what many urban young had done since time immemorial: playing, working, begging, thieving – and often simply causing a nuisance. Children were undoubtedly unavoidable, a result of population growth (and an overall population, one-third of which was aged 14 or under), and protective legislation which drove ever more children out of the workplace and into the streets. Escaping from overcrowded homes, large numbers of children resorted to the companionship of their friends and peers, in knots, gangs, or marauding noisy bands, to a degree that contemporaries simply could not recall. Many of them fell victim to the new policemen, armed with the acts from the 1820s empowering them to sweep up suspicious people. There thus emerged a widespread belief that there was a juvenile crime problem in the last years of our period.[23] But, like the evidence about drunkenness, much of this data was a function of new laws and of a new policing efficiency. None the less, there developed a distinct criminalization of a large area of urban life; the street games, organizations, and social life of children and young adults. They were, in large part, curbed and brought under more effective social control (a process which reached its apogee in the last quarter of the century with the coming of compulsory schooling). Nor was this a matter simply of local policemen stopping them; they were arrested, prosecuted, and imprisoned, in large numbers. One child in York spent a week in prison for playing cricket in the street.

By mid century, the sheer numbers of children and young adults criminalized and imprisoned in this way – not because of personal weakness but by the simple vagaries of changes in the law and law enforcement – forced Parliament to consider the wider problem of juvenile crime and punishment.[24] If the 'problem' of children's and youthful street crimes (or rather many of them) had effectively been purged – and driven into the new institutions – by mid century, it was perfectly clear that the more fundamental problem remained, of how to handle crime among the very young.

By the mid nineteenth century, one simple solution was to lock away the wrong-doer. The prison or reformatory, like the lunatic asylum and workhouse, were key institutions (bequeathed, of course, to the present day) devised in the first half of the nineteenth century both to remove society's various problems, and also to offer the prospects of reform. Bizarre as it may appear in retrospect, the early Victorian prison was inspired by a mixture of humanitarian and utilitarian ideals, designed to punish and to improve. So large scale had the prison population become by mid century that it was also clear that the prisons served to criminalize (and brutalize) large numbers of people, imprisoned (like the young) not so much for wrong-doings but for falling foul of a new, more efficient and draconian policing and penal system. Whereas at the beginning of our period the prison population was very small, a century later it had risen to a daily total of almost 29,000.[25]

Prison – like the workhouse – was idealized, by its proponents from the 1830s, as an instrument of social discipline. There is no question that the internal administration of prisons improved, thanks in large part to the work of a small band of humanitarians who exposed the scandals of unreformed prisons. Of course, the humanitarians' prison reforms were only one small part of their efforts (which ranged from anti-slavery to cruelty to animals) in order to civilize the English people – or rather, the common people. For our purposes, however, it is important to stress that they chose to do this through their own informal organizations, but also through establishing or encouraging new laws and institutions. The evidence about their motives is, inevitably, mixed, but on the whole there is little reason to doubt the humanity which formed the centre of their ideology; the conviction to end a range of English cruelties and sufferings – whether against the mad, the criminal or the poor. But, in all of these cases, it was felt that the closed institution would provide a crucible from which would spill forth a chastened and newly-reformed breed of people, more able to play an accepted role in society at large. Such ideals may now seem either naïve or cynical; a mask for more sinister purposes.[26]

Whatever the inspiration behind them, there is no doubt that prisons swept up large numbers of urban dwellers (rural people, too, of course) whose 'crimes' were trivial and would, in an earlier generation, have gone unpunished. From the 1830s onwards, there was a new and reasonably efficient system of policing and punishing the urban English. Whether it was a system that addressed itself to the real problems of urban life is open to doubt. But it was a system that had an undeniable impact on the nature of town life; in the conduct of life in the streets of urban England during the day and the night. In the case of the young, the activities which they had indulged in simply to survive in a hostile urban environment could now attract legal punishment; so, too, would many of their games and recreations. Along with the more striking outbursts of urban violence, the daily and apparently endemic troubles of street life henceforth found themselves under close scrutiny and control. Clearly, the police patrols, the magistrates' courts and the prisons could not purge the cities of all their noisy tumults or troublesome individuals. None the less, their success in purging urban England of a range of its taxing human activities was considerable. There was a marked subjugation of wide sectors of urban life by these new legal agencies. After all, that is what they were designed for.

The imposition of firm government was not solely a consequence of new policing. The railway and telegraph systems were also extremely important. The telegraph system of the 1840s provided national government with swift knowledge of troubles throughout the country; police and troops could be swiftly deployed along the railway system. In July 1839, when large Chartist crowds were thought to threaten stability in Birmingham, the *Annual Register* recorded

> The borough magistrates, however, who had for some days been in constant communication with the home office, had by this time bespoken a picked body of sixty policemen from the metropolis. The railway train delivered them at Birmingham that evening.[27]

Later that year, faced by the Newport rising, the Home Office informed the local mayor of the dispatch of troops and weapons. Some would march to Bristol, 'and can, of course, if necessary cross over by steamer to Newport the same evening'. Others, coming from Woolwich 'will proceed by railroad to Twyford, from there by forced march to Bristol. . . '.[28] Chartism, itself a localized and, therefore, a diffused movement, was faced by a resistant government, able to keep in step immediately with every Chartist move, to counter it as appropriate and, when physical trouble erupted, bring to bear the efficient and harsh rule of policing – thanks to the telegraph and train. Indeed, the government use of the railways was to be a major counter-revolutionary force throughout Europe in subsequent years (notably in 1848 and 1870–1).

Major unrest could not, however, be utterly eradicated, if only because its major inspiration – economic and industrial distress – continued to plague English life. In the summer of 1842, another wave of industrial unrest swept through the manufacturing areas, beginning among the Staffordshire miners but having its centre of gravity in the northern textile industries – the very areas which succoured Chartism. According to the *Leeds Mercury*, 1842 was 'The Chartist Insurrection'. Wage cuts, strikes, mass meetings, attacks on plant (the 'Plug Riots') all cumulatively provided a fertile soil for revived local Chartist activity. Time and again, police and troops clashed with strikers and Chartist leaders. And, just as the forces of the law were able, in general, physically to overawe the crowds, their work was completed by the courts. The police once more proved that they could break the heads of their opponents; the courts broke their spirits.

That motor of unrest, economic distress, began to revive collective complaints in 1847. 1848, the year of European revolutions, saw the last of mass Chartist activity. And the response of government in that year is symptomatic of their determination to marshall newly-created forces to maintain order. Troops ringed London, special constables were mobilized, artillery prepared, and 4000 London policemen positioned to combat the 1848 Chartist march. Most remarkable of all, more than 125,000 special constables were recruited, mainly, but not uniquely, from the middle classes. A similar story unfolded in a number of provincial cities. There were, it is true, violent eddies later in 1848, but that year clearly marked a turning point in the ability of government to cope with whatever threats of mass demonstration or violence came their way, through the astute use of police forces. True, certain violent instincts persisted in later years, but massive urban turbulence quickly receded as a major problem taxing English governments, and by 1851 it was widely agreed that the English had become a peaceable people. Indeed, the Great Exhibition of that year was so successful in large part because the armies of people who attended did so, 'not only without disorder, but almost without crime'.[29]

The chemistry of this social transformation is extraordinarily complex, and was a result not merely of the physical dominance of the force of government. No less crucial, though less obvious, was the way urban people had tamed themselves by channelling their energies and ambitions into organizations that sought to come to terms with, rather than challenge, the changes in society at large. By mid century, a distinct complacency had entered British commentaries on their new-found social tranquillity. It was as if people had come to terms with their own changed environment; had come to accept their lot and fate in life. The tranquillity of mid century England was, in many respects, deceptive. There were any number of urban disturbances much later in the century but they were not on the scale, nor were they as persistent a problem, so common in these years. Indeed, the mid – and late – Victorians were regularly congratulating themselves on having emerged from these worrying and uncertain years into a period of relative social and urban tranquillity – notwithstanding the continuing fears posed by 'unknown England' in the darker recesses of urban life. England had to wait till the summer of 1981 for a repeat of the urban violence and loss of urban control of the kind so commonplace in these years. However deceptive the mood of mid Victorian urban tranquillity, there could be no denying the remarkable contrast when set against that contrary experience which had dominated much of our period. The ungovernable people of England, now a nation of town dwellers, had become eminently peaceable and governable.

Part Three

Continuities

It is relatively easy to plot the changes in English life in this period, but much more difficult to assess the continuities in social behaviour. Moreover, by concentrating on those undeniable and major changes which so overawed contemporaries it is understandable that there has been a tendency to overlook or minimize those forms of behaviour which seemed relatively untouched by the process of change. Furthermore, the situation is made more complicated by the continuing debate among historians. There can be no doubt for instance that the popular recreations of the common people underwent a series of profound changes in these years. Traditional leisure pursuits were changed, banned or driven into marginality by a variety of (often local) forces: changes in landownership or use, urban regulations, newer occupational disciplines, and the political efforts of various interest groups. Yet it is remarkable that despite the undeniable changes, the English continued at the mid nineteenth century to enjoy many of the collective pleasures which would have been recognizable to their forebears. Similarly, the array of personal or communal ceremonies and rituals which characterized early Victorian life had origins that went back to the eighteenth century – and well beyond. Many of those rituals have been ignored because they seem so commonplace, so universal and ubiquitous a feature of English life over many centuries. But it is their very persistence and durability which catch the eye and which raise fundamental questions about the nature of social life in this period. So many rituals survived the upheavals attendant on urban change because they were rooted in the cycles and patterns of family life. And if there is any one social institution which weathered the changes better than others it must surely be the family. At the end of our period (and of course beyond) it remained – as it had been at the beginning – the fundamental component of English society; the crucible from which was created the personal and aggregate qualities (and vices) of English social life. Indeed, it is likely that there were so many continuities in English social behaviour throughout these years because so many of them were ultimately rooted in family life and relationships.

9 *People's pleasures*

Long before the advance of urbanization there had existed a powerful and growing campaign, among men of influence and property, against many of the popular recreations and games of the English common people. Fear, dislike and hostility towards the pleasures of the common people were as old as fear of the common people themselves. The pre-industrial popular culture of fairs, communal violence, music, local festivities and the calendar of trade celebrations, had often, individually and collectively been regarded as a threat to stability and good order. There were, naturally enough, exceptions to such propertied distrust, notably when carnivals, *mardi gras* or charivari were tolerated and even encouraged as the permissible outlet for plebeian turbulence and boisterous collective pleasure-making.[1]

The great bulk of this popular culture was particularly local, generally parochial, however much it might be rooted in the calendar of the agricultural year, the church calendar or trade cycles. Thus, while the general outlines of popular culture were broadly familiar across England, it formed a rich mosaic of local peculiarities. In the years of this study, however, popular culture came under increasingly fierce attack both from its traditional opponents – and from new ones – quite apart from the corrosive effect of industrial and urban change upon the habits of the English people. Although it would be premature to claim that the distinctively modern games and pastimes, in which the English took such pride in the last quarter of the nineteenth century, were indisputably in existence by mid century, it is true that the embryonic form of modern popular culture was detectable. Despite a host of parochial, trade or local exceptions, the popular culture of the pre-modern world was unquestionably in a state of flux in these years, more especially after 1800.

In many respects this is hardly surprising. Since, as we have seen, the combined effects of urban and industrial change were recasting most social attributes of the urban English, it was natural that their leisure pursuits would also change. The routines of urban life and the rhythms of industrial work demanded quite distinct personal qualities; they also came to impose new constraints on the pastimes of the English people.

People's leisures were largely shaped by economic and social circumstances. The ability of the great bulk of the population to pursue and enjoy particular leisures was severely limited by the constraints of economic life. The small propertied élite whose lives were materially secure could afford the ample time and expense of those costly pleasures which characterized their social life. Indeed it is significant that they were known as 'the leisured class'. For the rest – the great majority – life in general was consumed by the need to scratch a living as best they could. None the less the annual routines of life and labour were punctuated by pleasurable releases; religious festivals (Christmas and Easter which were in effect national holidays), feast days, local festivals, fairs and saints days. There were, too, the traditional agricultural breaks, more especially local celebrations at harvest and crop time, which are features of agricultural life the world over.

It would be wrong, however, to draw too rigid a line between work and leisure in the pre-industrial world. People able to dictate the pace and nature of their work could create free time for themselves; they could often take time off work – an hour or a day – in the knowledge that they could make up for lost time by extra effort later. This habit was, of course, one of the major obstacles and complaints of the first generation of industrialists anxious to inculcate new habits of persistent and unremitting toil in their mechanical enterprises. Similarly, in the pre-industrial world it is difficult and sometimes impossible to divide popular leisures between the rural and urban world. Indeed it is symptomatic of the encroachment of modern urban life that such a division came into being. As we have seen, London apart, all mid eighteenth-century English towns were small by modern standards. Geographically compact, they were linked by trade and services to the rural hinterland which was within easy walking distance for the town dweller. The movement of country people into the towns, on market days, for fairs, assizes, to watch and enjoy local festivals and customs, guaranteed a regular and periodic ebb and flow of contacts between urban and rural people, and ensured an intermingling of their varied pleasures.

The history of leisure had already experienced marked transformations long before the era of major urban growth. This was especially true in the sixteenth and seventeenth centuries when a dominant puritan ideology sought to purge the nation of many of its unacceptable pleasures. Theatre and a range of coarser enjoyments were effectively banned or removed from the calendar of local customs. A similar purification was directed at rural pleasures and especially at the ale-house (which was the main centre and locale for a wide range of popular leisures).[2] This was not merely a religious campaign – the ideology of a religious and fanatical minority – but was also economically inspired. The curbs on popular entertainments were also designed to clamp a firmer and wider labour discipline on labouring people addicted to the leisure customs of generations past. The eventual outcome was the creation of the English sabbath – a day set aside from working days and to be devoted to worship. Naturally enough this ideal was not universally attained and the battle by the devout to keep holy the Lord's Day continued to be fought throughout the eighteenth and nineteenth centuries. Victorian Sabbatarians were, however, not innovators but were continuing, in greatly changed circumstances, the traditions of earlier generations of godly Englishmen, anxious to keep pure the sabbath and to prevent its desecration, as they saw it, by the robust or boozy leisures of the common people.

Such restraints however were no impediment to the prosperous, who were growing in numbers with the economic advance of the nation, and for whom there developed an increasingly varied round of pleasurable pursuits, during the London 'season' and at the spa towns that proliferated in the seventeenth and eighteenth centuries.[3] At Bath, Tunbridge Wells, Scarborough, Harrogate and elsewhere there was available a highly sophisticated calendar of fashionable, and therefore costly, pleasures and rituals. The great bulk of the people could only watch and envy their betters as they cavorted through their expensive pleasures. Local people could enjoy certain natural facilities – at Scarborough for example – but by and large the common people were remote from the pleasures of their betters for the same reason that separated them in all other walks of life – the harsh economic facts of contemporary life.

In the light of economic circumstances, and the fierce opposition of certain sectors of propertied society, it is remarkable how tenaciously the common people clung to their traditional enjoyments. Indeed, the fiercer the attack, the more solid the popular resistance to curbs on their traditional recreations. As we might expect, the criticism of popular cultural forms were often presented in an indirect and oblique form: that they were a brake on economic (and therefore national) progress and prosperity. Not for the last time, vested interest groups were able to wed morality with economic utility; to argue that the nation would be a better and a more prosperous place by heeding their arguments.

The attacks on popular and plebeian pleasures gained force from the late eighteenth century onwards, as a new class of men – of property, substance and espousing a sensibility which demanded of others a new social discipline – gradually imposed their control on the conduct and politics of local society. In common with many other features of urban life, the changes that began to make themselves felt in English recreations were distinctively local. Cumulatively, however, they added up to a process of national change. Moreover, they were changes which affected rural no less than urban society. But it was in the towns that recreational change seemed most imperative, and nowhere more so than in the realm of blood sports and violent recreations.

It is seductively simple to caricature popular enjoyments in the pre-industrial world; to imagine that they were uniquely or overwhelmingly violent and gory. To a marked degree, such violent enjoyments are best remembered because they were so spectacular and because their after-effects – of injury, damage or death – are more likely to have been recorded. There was, of course, a host of other peaceful and unremarkable enjoyments. None the less, it was against the violence of popular recreations that polite society increasingly railed.

Bull-running and baiting, bear, cock, badger and dog baiting and fighting came to be viewed with growing dislike by increasing numbers of men of substance who viewed such pursuits as archaic, anti-social and corrosive of the social discipline so vital in a changing world. An end to blood and violent sports was viewed as a vital aspect of the 'refinement of manners' actively promoted to transform the social fabric of the English people.[4] Commonly associated with the evangelical movement, this process of refinement attracted more than the mere religiously committed. It appealed, for instance, to men who felt that many of the old pastimes were inimical to good social order, collective discipline and, increasingly, to the successful pursuit of economic life.

Certain objections were especially strong, few more so than the fear of large, boisterous and potentially uncontrollable crowds, some of the largest of which continued to gather for the public executions that remained among the most popular of all pre-modern public entertainments. Crowds, measured in their tens of thousands, were beyond the means of urban government to control and the sheer pressure of people frequently led to injury and death, to say nothing of encouraging further capital offences and widespread drunkenness. The traditional difficulties posed by the popularity of public executions were worsened, from the 1830s, by those railway companies that organized excursions to see executions. The crowds involved were simply enormous; often between 50,000 and 100,000 in London and the bigger provincial cities, and far surpassing the numbers flocking to other public entertainments, and perhaps even the most popular of political events.[5] There were, naturally enough, a variety of objections to public executions (not abolished until 1867), expressed most forcefully and persuasively by Dickens, but paramount was the problem of crowd control. Executions were a form of public carnival, and like all such carnivals they posed a persistent danger of spilling beyond the bounds of control. They were not unique in this, for there was a number of other popular pastimes which traditionally attracted large crowds. Shrove Tuesday football matches drew large audiences – to say nothing of the huge marauding armies of players intent on inflicting damage or injury on opponents. So too did the major boxing matches which, like executions, naturally attracted a criminal fraternity and excited heated passions among the spectators. Similarly, the major London fairs were able to attract large crowds at their periodic venues in and around the metropolis.

The most severely criticized and persecuted of all popular recreations in these years were those characterized by violence, cruelty and blood-letting. Objections to them were accentuated, from the 1780s onwards, by the rise of a powerful humanitarianism which sought, among other things, to reform English manners by curbing violence and cruelty.[6] The ranks of opposition were further augmented by the emergence of powerful local dissenting communities who not only came to exercise great economic sway in their urban localities but, as we have seen, were often able, from the mid 1830s, to exercise political power. Thus local regulations, social pressures, and private agencies, in conjunction with agitation within Parliament, were able gradually to impose restrictions on popular local recreations. And yet despite such efforts some of them survived the onslaught of propertied and legal opposition. Some were able to adapt themselves, to deflect the growing burden of criticism, but others inevitably succumbed. Throughout the first thirty years of the nineteenth century, complaints issued from a host of towns and cities – and London – that local blood sports continued to thrive despite a battery of local prohibitions, the determined opposition of private agencies and the new police forces. Many were, it is true, successfully suppressed, others simply faded away. But others remained dormant, resistant to the forces of suppression and were able periodically (though often furtively) to resurface. Blood sports and animal baiting were reported in a number of major towns well into the nineteenth century.[7]

It may seem strangely ironic that the attack on popular blood sports was not accompanied by a parallel restraint on the bloody and rural pastimes of the propertied classes. As Pamela Horn reminds us, 'For the landed classes of England, hunting, shooting and fishing were *the* major leisure activities'. In fact these recreations increased in popularity during the first half of the nineteenth century, nurtured by the growth of a prospering middle and upper class and facilitated by the new communication systems. Country estates established their own game departments and protected their recreational investments by employing growing numbers of keepers against the inevitable and understandable furtive attacks of hungry country people. And the whole process was shored up by a repressive legal framework (thirty-three new Game Laws were enacted between 1760 and 1816).[8] Poaching was, for many, a matter of family survival. For others it became an organized business, employing gangs prepared to do pitch battle with the keepers. Killings, serious injuries, executions, transportations and punitive jail sentences all characterized the struggle of the propertied orders to preserve their recreational interest against the more basic needs of poorer people.

The inconsistencies in the attitudes to blood sports were plain for all to see. In the words of the Rev. Sidney Smith: 'Any cruelty may be practised to gorge the stomachs of the rich, none to enliven the holidays of the poor.'

Time and again, similar sentiments were expressed by those who took a balanced view of the problems apparently posed by recreations. A Stamford magazine complained in 1819

> The privileged orders can be as cruel as they please and few are the mortals who dare say wrong they do; while every evil action of the lowly is trumpeted forth. . . .[9]

Such complaints of inconsistency had little effect and the legal defence of field sports was zealously pursued throughout the nineteenth century (in 1870 there were 10,580 game-law prosecutions in England and Wales).[10] Yet these were the same years which saw a propertied onslaught against the turbulent pastimes of the common people.

Almost as striking as the disappearance of certain recreations was the transformation and adaptation of others. This was particularly so in the case of football which, in a matter of a mere three generations, went from traditional folk game – often of considerable turbulence – to the recognizably modern sport which was disciplined, codified, rational and nationally uniform.

Notwithstanding this transformation there were many areas where the ancient games of football survived in their traditional form (notably as a youthful, male village custom of Shrove Tuesday). Indeed, a number of these games continue to this day. More significant, however, is the transformation of the game, primarily through the agency of the reformed public schools.[11] Yet the fact that these two cultural forms survived, side by side – the unreformed and the new, regulated urban game – raises questions about the broader development of popular recreations and culture in these years.

It has been commonly argued that the forces of urban expansion and industrial change were the key factors that purged the nation of its traditional recreations; that they were instrumental in creating a new popular culture, at the heart of which lies a sharp and growing distinction between work and leisure time. Although by the mid nineteenth century these changes were quite limited, there were none the less striking transformations in popular culture, brought about especially through the increasing restrictions on free time and physical space. It is an obvious point to make, but the development of urban areas – with their encroaching apparatus of local administration and law enforcement – often took place by absorbing open spaces and placing restrictions on much of the space that survived. Throughout the eighteenth century, games, fairs, festivals and recreations had, quite literally, normally taken place on common land to which local people assumed they had a customary right of access. Increasingly, however, that land was not only a target for urban growth and development (in the country it was under attack through enclosures) but forms of local popular culture were distrusted and attacked by propertied critics. This was largely because so many facets of popular culture were unsympathetic to the ideology of the emergent bourgeoisie. Clearly this was particularly true of blood and violent sports which often inflicted damage and injury on people and property. But it is also obvious that some pleasures, whether pre-industrial football or bull-running, were ill-suited to town life which was itself ever more regulated, controlled and disciplined and whose inhabitants were expected, by their social betters, to display similar virtues.

The attacks on popular culture were specifically local in character, but took place throughout England. Basically men of substance tried either to repress and drive out local popular culture, replacing it with something more 'suitable', more rational, more sympathetic to their perception of what was appropriate, or to encourage its demise by providing seductive alternatives. Athletics competed against folk football. Mechanics Institutes and reading rooms sought to seduce men from other traditional sports. And all the time the battle involved a perceptible shifting of popular culture away from its traditional communal roots – on local common land – to 'safe' areas where it would not disrupt local economic and social life, and where it could be conducted in a style more in keeping with the disciplined work ethic so precious to the employing classes – but so alien to their workforce.

Vast acreage of communal land simply disappeared along with miles of public footpaths, though often this was merely as an accidental by-product of urban growth. More often, however, it was part of a concerted drive to encourage the more efficient conduct of economic life and to alter the cultural pursuits of the labouring people in order to render them more malleable and accommodating. Whatever the cause, the effect had become everywhere noticeable by the early nineteenth century. Time and again contemporaries remarked on the decline of open space. Sir Edwin Chadwick noted in 1834 'In the rural districts, as well as in the vicinities of some of the towns, I have heard very strong representations of the mischiefs of the stoppage of footpaths and ancient walks, as contributing, with the extensive and indiscriminate inclosure of commons which were play-grounds, to drive the labouring classes to the public-house.'[12] Paralleling the attack on public access to the old common lands was an ever more stringent municipal control over the streets of urban England. Public nuisances on the street were, it is true, a traditional complaint of the propertied orders. But the development of more effective local government after 1835 and the agency of the new police forces provided both the will and the means of controlling urban streets. This was noticeable, for instance, through the attacks on children's games which, of necessity for the poor, were invariably conducted on the streets. Now they could be banned and the childish or youthful players barked at, cuffed, and even arrested.

At the very time popular culture was under attack there developed a veritable renaissance of middle- and upper-class cultural pursuits, with all those physical artefacts – of museums, libraries, galleries and meeting places – which became such prominent features of urban England. This was all in sharp contrast to the increasing material impoverishment of plebeian culture. It might be felt that this is simply a particular feature of a more general process; of rising, propertied expectations paralleling the growing immiseration of the lower orders. It was not, however, a universally bleak picture, for what is also striking is how popular culture often managed to survive (though often in transmuted form) *despite* harsh circumstances. Nowhere were traditions more clearly maintained than in the ale-house.

There was an uninterrupted descant of complaint against the place of drink and how it continued to entice generations of working men. In some respects the pub's role was enhanced in these years because, as the public domain of popular culture was increasingly attacked, the place of drink came to occupy a proportionately more significant role. It would be wrong to consider the pub merely as a place for drink, for it had traditionally provided a range of social functions. It was the provider of a wide range of services and a meeting place for a host of local organizations and institutions, and these functions were enhanced by the process of urban growth. It was a common complaint or remark that the only available local meeting place was the church or the pub; 'the former was seldom open, while the latter was seldom closed'.[13] There were few plebeian organizations – political or social – which did not meet in local pubs. The pub provided warmth, light and companionship in a convivial atmosphere – social qualities all too often missing in the bleak domestic environment. In the words of Peter Bailey,

> in an age of social dislocation the pub remained a centre of warmth, light and sociability for the urban poor, a haven from the filth and meanness of inadequate and congested housing. A magnet for the disoriented newcomer and the disgruntled regular alike.[14]

It is symptomatic that from the pub there emerged that symbol and instrument of working-class popular culture – the music hall – later in the century.

Two factors, mutually reinforcing each other, determined that drink and drunkenness would remain a major problem in plebeian life: the centrality of the pub itself in working-class communities, and the continuing (and in some respects worsening) nature of domestic conditions. Drunkenness was not peculiar to the lower orders of course. Throughout the eighteenth century excessive drinking in propertied and educated society was an unexceptional norm. The vernacular expression, after all, is 'Drunk as a Lord' not 'Drunk as a labourer'. However, it was concern about plebeian drunkenness that dominated eighteenth- and nineteenth-century debates about contemporary drunkenness. It was a concern that could be supported by some startling evidence about drunkenness and alcoholic consumption, through a series of alarming statistics. Of course we need to recall that from the 1830s the new policemen began arresting large numbers of people for a string of minor offences – including drunkenness – previously unchecked. Not surprisingly, drunkenness in towns seemed to increase dramatically.[15]

A growing number of informed observers felt that the repression of plebeian pleasures, more especially by the Sabbatarians, was ironically driving the poor to drink. As ever more pleasures were restricted or purged, growing numbers of urban people found the pub their only place of enjoyment. Engels remarked of Manchester in 1844 'Liquor is almost their only source of pleasure, and all things conspire to make it accessible to them.'[16] Yet the natural alternatives – the obvious local places of pleasure – whether gardens, parks or museums remained securely closed, thanks to pressure of inflexible opponents of popular culture.

There were, it is true, determined efforts by some, notably the evangelicals, to reform and enlarge plebeian popular culture, but their methods and aims were generally unrealistic and unattractive. 'Rational recreations', games under paternal supervision and the direction of social superiors, and the evangelical promptings to pursue self-improving domestic entertainments were, understandably, utterly unattractive. Domestic pleasures, so much a feature of propertied life, could be created and the attractions of 'unseemly' pleasures (notably the pub) be countered only by a marked improvement in the conditions of domestic life itself. In fact it was the unrelenting bleakness of the home which drove so many people out of doors for their pleasures. In retrospect it is now easy to see that this could change only via significant material and economic improvement for working people.

Throughout these years there was an increasing attempt in public and private to inculcate new social qualities among working people; to encourage the virtues of disciplined, orderly and 'rational' enjoyments in contrast to the more passionate and boisterous features of plebeian life. Private attempts were made to organize local cultural attractions which would both entertain and 'improve'. Later still, even national government determined to set aside money and facilities for public spaces. But it was local and often charitable efforts which, from the 1850s onwards, came to make significant provision of urban space for public use.[17] Certainly in the years up to 1851 the establishment of parks and gardens in no way compensated for the contrary process, long in train, of removing forever common land from public use.

Rational recreationists placed great value – and hope – on the power of music to transform popular culture. In this they could build upon the wide popularity of a number of musical activities. Indeed there were few areas of popular culture which were so universal, traditional and all-embracing as the English commitment to music, dance and song. In these years, moreover, the popularity of music increased. Choral and communal singing, for instance, was given a major boost in many working-class communities by the Methodist commitment to choral singing and the consequent proliferation and perfection of popular musical skills. Indeed wherever music became a dominant feature of local plebeian life, whether choral, bands or festivals, the key determining factor seems to have been the presence of a strong local dissenting (especially Methodist) tradition.[18]

The image of working-class musicality in the nineteenth century (and the twentieth) is often portrayed in mythical images, more fitted to fiction than historical reconstruction. Many commentators have been seduced into viewing plebeian musicality as the embodiment of all that was (and is) good about the independent, self-directed virtues of plebeian life. Whatever the judgement, contemporary observers and modern historians generally agree that musicality was widespread in working-class life. As early as 1788 Charles Dibdin wrote (of Halifax),

> the facility with which the common people join together throughout the greatest part of Yorkshire and Lancashire in every species of choral music, is truly astonishing.

Thirty years later, much the same point was reiterated: 'the spirit of music pervades the people of this district in a manner unknown and unfelt in the rest of our island'. Much of this music was domestic; families and friends using their limited leisure time to great musical effect – singing and playing the fiddles and flutes which were the popular instruments of pre-modern musical culture. Few working people could match the weavers' musical traditions which were handed down to successive generations both within the family and the trade. It was a tradition which even caught the attention of Parliamentary observers in 1841. Weavers in Lancashire and Yorkshire were described as

> famed for their acquaintance with the great works of Handel and Hayden, with the part-music of the old English school, and those admirable old English songs, the music of which it is desirable to restore to common use.[19]

Religious life was clearly seminal in encouraging and spreading popular musicality; the relative backwardness of music in the Church of England was accepted as one of that church's disadvantages. One man argued in 1790, 'It is not rashness to assert that for one who has been drawn away from the Established Church by preaching, ten have been induced by music.' As a result, a great deal of popular musicality was sacred; none more important than Handel and the 'Messiah'. In Halifax it was first performed in 1766 and soon became – and remains – an extraordinary northern cult. In the winter of 1768–9 it was performed on fifteen consecutive Friday nights at Holbeck Chapel in Leeds.[20] By the mid nineteenth century, it had become a major ritual of the Christmas period throughout the urban north. As local working-class institutions proliferated – Sunday schools, choral societies, bands, and chapels – they vied with each other to present the 'Messiah'.

By the mid nineteenth century, all forms of musical activity had greatly increased in the towns, helped in large part by the technical innovations in printing and instrument manufacture which, in conjunction with simpler distribution systems, greatly reduced the cost of sheet music and musical instruments. Furthermore, working-class musicians were now able to purchase their relatively costly instruments thanks to embryonic local hire-purchase agreements. Here, as in other areas of plebeian cultural life, their social betters strove to shape and direct this traditional and abiding working-class cultural form into the channels best suited to the rational disciplining of urban society. Manufacturers and employers encouraged – and often subsidised – bands and choirs among employees; local dignitaries organized concerts and competitions. Much of the music was sacred, enjoyed on its own terms but positively promoted by social reformers in the hope that it would both entertain and elevate (it was felt that sacred music would inspire audiences and musicians to contemplate a united rather than a socially divided society).[21]

It would be quite wrong to imagine that the 'respectability' of so much popular music was a function solely of an ideological middle class determined to re-shape the interests and values of their inferiors. This musical culture – which really flowered in the second half of the nineteenth century – also grew from and reinforced the independent world of plebeian respectability and self-improvement that was itself so pronounced a feature of working-class life. Working-class respectability was not merely a quality in the eyes of the middle-class beholder (then or now) but was a powerful, independent and determining feature of working-class life. It shaped the social style of generations of working people; it created for instance the autodidact (so loved of many historians who none the less scoff at other dimensions of this same quality), and it created those generations of working-class musicians who came to represent one of the most powerful and persistent cultural forms of plebeian life. None the less, at the same time it is indisputably true that music was regarded as inspirational and was consequently promoted among working people by their betters.[22]

Music offers a particularly good illustration of the continuities in popular culture, from pre-industrial to modern society. But even music underwent a process of modernization consequent on the wider changes in industrial and urban life. And a similar story can be told of many other cultural forms. It was true, for instance, of the habit of 'taking the waters'. The spas of eighteenth-century England had, as we have seen, become a mecca for people not merely interested in the cures allegedly bestowed by local waters, but also in that range of social activities provided in those burgeoning towns. Bath, the epitome and peak of fashionable spa life, was denounced by Charles Wesley as 'the headquarters of Satan', but in fact its social tone was less overtly coarse than its opponents alleged. Its popularity, however, was undisputed and was reflected in its population which grew in the century to 1800 from 2000 to 34,000.[23] Bath clearly led the way, but there was a sprinkling of other spas, with their own array of recreational activities for the propertied and the infirm right across the country – and of course in London itself. There, as in Bath, were to be found the prosperous, the *arrivistes* and the ambitious, besporting themselves as local taste dictated and their pockets allowed.

With the exception of local people, the poorer sorts were, of course, unable to enjoy such delights. The great bulk of working people had little free time, spare cash or adequate opportunities to enjoy what were, in fact, costly commercial pleasures. Similarly the cost – and time – of travel precluded long-distance migrations by the poor. Many of them could, however, travel (generally by foot) to the seaside. Swimming was a traditional enjoyment wherever the opportunity arose – in rivers, lakes or sea – but what transformed this recreation into a major and distinctively British phenomenon was a switch in medical opinion in the late eighteenth century. Led by Dr Russell of Brighton, doctors began to argue that sea water contained those medical qualities they had previously claimed for spa waters. Their patients thus began to travel to the sea – aided by improved coaching services and roads to the south coast or by new shipping lines down the Thames estuary. There, just as they had done at the spas, they drank and bathed in the local water. Scarborough, for instance, enjoyed the unique advantage of both spa and sea water and was able to make a painless transition from spa to seaside resort. In general, however, the new seaside urban areas that rapidly developed around this peculiar custom began to lure their clientele away from the inland spas. They were able to do this by offering replicas of the facilities available at the spas – the hotels, apartments, gaming and assembly rooms, musical soirées, library and dining facilities and the like – in addition to the growing and national appeal of the sea and the coastline. A number of the southern resorts, notably Brighton and Weymouth, received the incalculable boost of royal and aristocratic patronage.

Architecturally and socially, the new seaside resorts consciously modelled themselves on the spas, although there developed marked social distinctions between the resorts. And these distinctions were to be sharpened as the nineteenth century advanced by the impact of the railways, which began, from the 1840s, to deposit periodic waves of working people at the seaside. Even before then, certain resorts in the south had come to reflect the middle-class – as opposed to aristocratic – tastes of their clientele.

Later, others were likewise to become plebeian, or 'vulgar'. In the process the seaside resorts had, by mid century, begun to register different and distinct social tones and styles. Some, like Scarborough, managed to cater for and to reflect the different tones of their separate social groups. Of course, this was a process made possible by the railway revolution and the economic changes that enabled growing numbers of working people, organized through their Sunday schools, factories or other institutions, to afford a trip to the coast. Whatever the cause, the outcome was undisputed: the development of major urban areas on the coast whose prime *rationale* was the provision of seasonal pleasures, or even longer-term retirement and health cures.[24]

The importance of the resorts can be measured in their physical growth. Seaside towns registered the largest increase of any type of town in the first half of the nineteenth century. Between 1801 and 1851 the eleven largest resorts recorded a population growth of an astonishing 214 per cent. Even the growth of the industrial towns came nowhere near this figure. The resorts were not of course as big as the major cities – the biggest, Brighton, had a population of 65,569 in 1851 – but their *rate* of growth was significantly greater. Equally interesting is the fact that, of the nine largest resorts in 1851, only three (Scarborough 12,915, Whitby 4765 and Blackpool *c.* 2000) were in the north.[25]

To a substantial degree, this growth was accelerated by the coming of the railways, which transformed many resorts from isolated regional spots, into major and nationally attractive towns. Earlier, Horace Walpole had remarked that 'One would think that the English were ducks, they are for ever waddling to the waters.' The trains reinforced this national urge. Dr Granville, chronicler of the early nineteenth-century spas, felt that if railway development continued, 'the whole nation, at length, will be on the move'.[26] The ever-extending railway system linked more and more inland areas to coastal ports, villages, remote hamlets, and even created some seaside resorts *en route* to other destinations; Rhyll and Bournemouth for instance. And throughout the 1840s the railway companies captured the working-class travelling market by providing excursion trains to the coast (and countryside). Wherever a new railway line opened a link to a coastal town the pattern was similar; hoards of working people from the industrial hinterland descended on the coast in crowded excursion trains. It caused chagrin and open irritation among their social superiors who resented plebeian invasions of their previously exclusive pleasures. Blackpool was one such case and, according to the *Preston Pilot* in 1851,

> Unless immediate steps are taken, Blackpool as a resort for respectable visitors

will be ruined. . . . Unless the cheap trains are discontinued or some effective regulation made for the management of the thousands who visit the place, Blackpool property will be depreciated past recovery.

Opposition to excursion trains to Scarborough came from those with 'no wish for a greater influx of vagrants, and those who have no money to spend'. A month after the line was opened in 1845 the first excusion train arrived from Wakefield and Newcastle.[27]

At various points around the coast, once the railways made possible the *mass* enjoyment of the seaside resorts, entrepreneurs quickly followed to provide (unless banned) that array of material and transient pleasures working people soon came to expect at the seaside resorts. By mid century the broad outlines of popular (and more refined) seaside enjoyments were well established; the sea and the beaches with their bathing machines, the piers and the sailing, the varied classes of accommodation and dining facilities, local amusements – permanent or temporary – a veritable galaxy of cheap attractions for poorer visitors. While it is true that at the older resorts more prosperous society sought to preserve its exclusive and refined facilities, pressure could normally be expected from plebeian or commercial demands. It was no accident that the Queen moved from Brighton to the Isle of Wight once the trains came to Brighton in 1841.

Thus there evolved the first phase of that modern British phenomenon, the seaside holiday, which, if only for the day, rapidly established itself as a national cultural form. Geography, it is true, was also instrumental; no inland town is more than about 70 miles from the sea. But it was the economic transformation of plebeian life, in conjunction with the railways, which, by the mid nineteenth century, enabled millions of working people to enjoy, however fleetingly and inadequately, the styles and tastes pioneered a century earlier by their social betters. It was, in a sense, one dimension of the democratization of recreation; a process that was to spread and accelerate in the last years of the nineteenth century. But, in the process, working people imposed on this particular feature of national culture – as they did on so many others – their own distinctive style and taste. The end result was that, beneath the apparent similarities of national recreations (in this case the pleasures afforded by a trip to the coast) these lay the remarkable fissures of English social class. English people by the mid nineteenth century enjoyed themselves – much as they lived and worked – within the sharp and apparently unbreachable divides of social class.

One striking feature of the trip to the seaside was that it was often a family affair. In this respect at least it was an unusual form of recreation. Indeed, it is striking how many recreations were divided by sex. Most of those dealt with here were specifically or overwhelmingly male, though it is easy to see why this should have been so. Public recreations and pleasures were overwhelmingly male dominated. Women of the propertied orders enjoyed on equal terms many of their social group's pleasures – music, soirées, balls, masques and the endless rounds of entertainment and dining. But the better-known participatory sports were – and remained – male dominated. This was in part merely the reflection of the prevailing views of what constituted manly or lady-like activities. It was axiomatic among many people of substance that sports and physical recreations were manly and therefore unlady-like. This was especially true of the more plebeian sports. Yet even in fashionable circles, women's leisures were by and large generally determined by the constraints – both physical and ideological – of domestic life. The more prosperous the home, the freer the woman of the house – and her daughters – to develop a recreational role within the home. This was also true – at a qualitatively different level of material existence – lower down the social scale. For poor women, bowed down by work and child-rearing in the starkest of material and domestic circumstances, there was, all too often, precious little time or opportunity for leisure. When it arose it was generally with and among their female peers, sharing spare moments in conversation and mutual self-help. Indeed, it seems likely that it was from this particularly restrictive recreational world that the commonplace myth arose of female 'gossip'. In fact, meeting and talking together often offered one of the few recreational outlets for generations of working-class women who, because of domestic circumstances, were not even able to enjoy the meagre public pleasures of their menfolk. Thus it is likely that despite the vast gulf of material circumstances between the social classes, the most determining factor in the recreational lives of women was the nature of their particular domestic circumstances.

The railway and the development of the seaside resorts were then especially important for poorer women. It is, of course, difficult to know how far down the social scale these benefits percolated for, even in their cheapest form, a railway excursion and a day at the coast were clearly beyond the reach of many of the urban poor. And this was to remain so throughout the nineteenth century. Set against this was the fact that a great deal of philanthropic effort, through churches and factories, went into organizing seaside trips for working people by the 1840s. This may, of course, have merely whetted the appetites of people unable normally to enjoy such treats; it may indeed have compounded frustrations about the inaccessibility of public recreations. None the less, by mid century the taste for the seaside trip had become an undeniable characteristic of urban working people; a taste only to be satisfied by regular (normally annual) trips to the coast. This was not, obviously, a natural, lemming-like instinct, but more a culturally determined pleasure – the enjoyment of which was highlighted by the mundane and unpleasant starkness of daily urban life.

Much of the charitable effort to take urban people to the coast was to provide for children. Partly to take children away, briefly, from the material dangers and squalor of urban, mainly industrial life, schools, churches and voluntary organizations progressively came to ferry thousands of children to the seaside. Indeed it seems likely that it was among the young that the appetite for the seaside was first created – and whetted. Among children of all social classes, the coast – the beach, the sea, the coves and rocks – became not merely enjoyable treats in themselves, but part of a widening social education.

The recreations of the young in these years illustrate the dangers in offering sweeping generalizations about the transformation of recreations. Among children – of all social classes – the traditional recreations remained remarkably durable and resistant to change.[28] The sporting ethic of tough masculine team games, so much a feature of late Victorian England, was shaped and perfected within the reformed public schools. With the coming of widespread schooling many of those reformed games – football, cricket, athletics – were introduced into poorer communities. Yet there was another cultural tradition – of continuity – which ensured that many local games survived in their pre-industrial form. But the really important recreational culture among the young was the informal, spontaneous and the traditional; those games, songs, verses and rhymes that were passed from one generation to another. Seasonal games dictated by the weather or, in the country, by the agricultural cycles, appeared, died away and reappeared as naturally as the seasons themselves. Rural life afforded an abundance of outdoor activity – of fishing, hunting, food and egg collecting, climbing and hiding, swimming, and cruelty to animals – which had been a perennial feature of rural children's lives.

In the urban areas such enjoyments were obviously limited, though not as completely inaccessible as we might imagine. It was a widespread complaint, as we have seen, that urban children (like their parents) were denied access to open spaces and parks. Inevitably then, it was on the streets, alleys and pavements that growing bands of children were obliged to seek their recreational enjoyments. Indeed it was the sheer physical presence of the young, with their attendant noise, games and general indiscipline in urban areas that so irritated and annoyed those adults whose own children were able to enjoy their games and pastimes within the comfortable circumstances of the home. The more prosperous the home, the more likely it was that special provision would be made for children; nurseries, playrooms, individual bedrooms, nannies and servants, and that expanding array of material toys which became so ubiquitous in western Europe and North American homes. For the poor there were few such luxuries. Their homes swarmed with children and when the weather improved home life was made a little more tolerable because the children poured on to the streets for their fun: 'whenever the rain is not actually pouring down in torrents, they turn their children out to find amusement and subsistance on the streets'.[29]

Among such children were to be found the timeless games and pleasures, individual or collective, which were graphically captured by Henry Mayhew in his accounts of mid nineteenth-century London. Many of the games required no play-things. Others did, however, and it is significant that there were toy manufacturers catering even for the very poorest of urban children. 'Bristol' toys, for instance, 'carts, horses, omnibuses, chaises, steamers, and such like – nearly all wheel-toys' were 'the common toys made for the children of the poor, and generally retailed at a penny'. As one toy-maker said to Mayhew, 'As all my goods go to the poor, and are a sort of luxury to the children, I can tell what's up with working and poor people by the state of my trade.'[30] But these and the embarrassment of play-things enjoyed by the more prosperous were, in some respects, incidental to the main recreational culture of children. Stretching back for centuries there was, quite simply, a folk culture of childhood which seems, as far as we can judge, to have remained relatively unaffected by the process of urban change. Indeed, many of those songs, games and verses survive in local or modernized form to this day.

The survival of children's recreational culture has a much wider historical significance. How are we to explain the continuity of this recreational culture in a period when so many other features of popular culture were in a process of transformation? At one level it is a cautionary corrective to the notion of a sharp divide between modern and pre-modern cultural forms; between urban and rural worlds. And it reinforces the degree to which the new urban areas absorbed so many aspects of pre-modern culture. This was true in patterns of death and bereavement, of civic rituals and certain adult recreations. Moreover, it is also a reminder of the efforts made by contemporaries to maintain and to cling to significant cultural beliefs and experiences – even in the depressing material circumstances of the towns and cities. Recreational forms were, for a great variety of reasons, aspects of a popular culture (both high and low) which was vigorously defended and adhered to. Some were destroyed and purged by the accidental consequences of urban life; some by the deliberate pressures to reform and refine the lives of local working people. Yet it is the continuities – even in extremely transmuted form – which are as striking as the breaks.

10 *Rituals and ceremonies*

Rituals are so basic to individual, family and communal life that it may seem barely worth the trouble of discussing them. Because they are so commonplace they tend to be ignored, to be taken for granted as part of the fabric of everyday life. But in these years of transition, when the nature of daily labouring and social life was so transformed in the crucible of urban society, it is to be expected that contemporary rituals would similarly change. Any historian working in a new university ought to be readily aware of the role and significance of rituals. The elaborate ceremonies that punctuate the calendar of those institutions were, like their very physical form, called into being at a particular moment in time to provide legitimacy, sanction and pomp and to add to their modern proceedings the trappings and dress of more ancient academic traditions. These bizarre and colourful inventions would have been attractive to early nineteenth-century city politicians, who similarly surrounded themselves with rituals. Urban ritual had, of course, been an important feature of city life since time immemorial, its role not merely sanctioned by longevity and custom but also hallowed because it was so important a part of the nature and governance of urban life. Those rituals that provided sanction and authority for the transfer, acceptance and display of urban power were paralleled by a multitude of ceremonies associated with other public or private bodies and institutions from the law to humble trade guilds. Of course it is true that the rural world was similarly characterized by its own distinctive ceremonials, reflecting the seasons and patterns of agricultural life, village customs or church celebrations. But towns and cities became more strikingly committed to such rituals if only because their ceremonies were so numerous, better attended and more colourful. And the appeal of urban rituals was so powerful that they attracted thousands of people, spectators and by-standers, from throughout the rural hinterland. Indeed the trek to the town or city to witness the local periodic ceremonies, and to enjoy the mixture of fleshy delights which generally accompanied them, became ever more prominent with the improvement in transportation.

Many of the pre-industrial communal ceremonies were related to, and often a by-product of, contemporary economic life. Ceremonies marking the change of seasons or the completion of tasks, for instance, alongside the obvious and rational ceremonies of the religious calendar provided even the humblest with cause and occasion for enjoyment.

Fairs in villages and towns had long provided a myriad of attractions for a whole range of people in rural life. Thousands flocked to neighbouring towns seeking work and hirings, to see relatives and friends, to enjoy the fun and spectacle of the travelling shows and artistes, and often to fall victim to the brandishments of the ale- or whorehouse, or even the recruiting sergeant. A year's wages could disappear in a week in this heady atmosphere of collective pleasure seeking.[1] Yet it is a sign of the transformation of English society that these and similar popular rituals came under the ever more effective attack of local men of substance who were anxious both to preserve local peace and decorum and to purge their localities of the surviving features of what they took to be a barbaric and backward-looking age. For men in local authority, the timeless plebeian rituals that found a focus in local fairs, wakes and recreations were attacked, controlled, outlawed and purged. Indeed it was the determination to break the cycle of popular rituals, and to impose a new, more orderly, controlled and supervised calendar of urban ritual that was partly responsible for the pacifying of urban life by the mid nineteenth century.

Ironically, however, the men who exercised power in the new urban areas were as anxious to create their own elaborate codes of urban ritual as they were to change those of their inferiors. Civic ritual was, of course, as old as towns themselves. Parades of local officials, the mayor and aldermen for instance, institutional ceremonies and civic religious services marked the beginning and end of the cycles of local government. As more and more of the new towns were incorporated after 1835 there was a parallel increase in the use of elaborate insignia of local office – of chains, mace, robes and processions. Such trappings of office were normally copied from older urban ceremonies. Now, however, they were the vestments of the struggle for political power which was waged between new and old economic and political groupings in the towns.

> Seats on the council, the aldermanic robe, and the mayoral chain became the yardstick for measuring success in the battle between a traditional Tory-Anglican establishment and a new Liberal-Dissenting economic life.[2]

It is significant that the new class of men who came to dominate and control local (though not national) government should appropriate to themselves the trappings and artefacts of high office. Despite their undoubted economic standing, they clearly felt that it was important to acquire, through the rituals and garb of a more traditional form of governance, both self-esteem and public acknowledgement. Such rituals, periodically paraded in public, conveyed legitimacy and approval in the complex social psychology of the wielding and acceptance of political power. At its most extreme such political rituals had evolved in the most minute and detailed form in the conduct of Parliamentary affairs, whose proceedings at times seemed positively hindered by the strict adherence to ancient ritualistic customs. Ultimately (though generally later than our period) the rituals of the new urban government found a home – a physical expression – in the gothic town halls that dominated the skyline of the late Victorian city centre.[3]

Both in the civic and private rituals of contemporary life, food and drink played a significant part. Feasting was, after all, a universal form of celebration. Indeed most of the nation's ceremonies, whether of a religious or secular nature, were celebrated by collective or family eating. There is, of course, nothing peculiarly English in this, for a host of societies, traditional and modern, use the rituals of slaughter, cooking and feasting to mark their own distinctive festivals. But we need to recall that in poorer societies – which include England in these years – a feast (as opposed to the mere satisfying of basic hunger) was in itself much more of an event than it might seem in a well-nourished prosperous western society. The provision of a varied menu, of food and drink and more especially of meat, formed a sharp contrast to the plain basic diet of most people's lives. Indeed the gift of exotic foods – a roast ox for instance – was a characteristically generous offer to the local poor. Thus local civic rituals were, as they had long been, accompanied by feastings; local elected civic officials confirmed their own newly-won rank by a large feast. And as the nineteenth century advanced, civic dinners punctuated the municipal calendar with increasing regularity. As travel became easier, and as visits of prominent people became more commonplace the civic banquet became a recurring and elaborate local celebration – its lavishness a token both of a town's well-being but, more especially, it was often a reflection of the status of local politicians.[4]

Politicians were not the only urban group to insist on eleborate displays of public rituals. Legal power, for instance, had traditionally been sanctioned by its rituals of judicial displays and processions. Throughout the country, local assizes were opened by colourful public ceremonials. The assizes were important at a number of levels; for national government to pass on its views to the localities and, in return, to listen – through the judges – to local propertied opinion.[5] But the whole process was accompanied (as it was in London) by an impressive and elaborate display of time-honoured rituals. Assize dinners, balls and sermons provided a rich tableau of local, yet nationwide, gatherings. They were also events of great political significance; a reminder, for those large crowds which gathered to watch the processions, of the power and ultimate fearfulness of the law. At its most awesome, assizes had the power of life and death. Indeed the rituals of executions were themselves important events in the localities and were carried out in public because they were thought to have a major influence on those armies of spectators who flocked to the gallows. Public executions were 'very much a popular spectacle, even a national institution', justified 'in the name of deterrence, of the vindication of the law, of the protection of the individual from arbitrary punishment'.[6] But above all else, this and other rituals of the law formed periodic highlights in the social life in the assize towns; rituals designed to impress upon everyone of the presence throughout the land of the agents of royal justice, and to overawe the observers with the splendours of the legal process.

Corporate and legal bodies were, however, only the most obvious and lavish of a multitude of organizations whose colourful rituals were paraded before a curious public, always keen to witness spectacular pageantries. As the capital city, London had long been accustomed to major civic and political rituals, and, of course, to those varied ceremonials associated with the seat of government and royalty. Royal ceremonies with the accompanying pomp of a military presence continued to provide a series of major spectacles throughout the nineteenth century. Many of these became personally associated with the Queen, and as Victoria aged, her various anniversaries were celebrated with ever more colourful and lavish tributes to the person of the monarch and, more colourfully still, to the human and racial variety of the expansive British empire. The people of London, and the legions of people able, thanks to an efficient railway system, to visit London to witness such pomp, could enjoy the apparent glories and achievements of monarchy, aristocracy and empire without leaving the security of their native shores. Moreover, however deprived and wretched a substantial part of the urban population (upwards of one-third throughout our period) there was a marked and undeniable popular enthusiasm for the rituals (and indeed for the very institutions) of pomp and ceremony, more especially surrounding the monarch. There was nothing new in this. The monarchy and aristocracy and their costly but colourful rituals, played out so close to – yet utterly distant from – the gaze of the capital's populace, had provided a traditional occasion for public entertainment. Sometimes, by way of contrast, however, royal ceremonies were the occasion for flurries of hostility to the royal or aristocratic personages. In 1795 George III was booed and greeted by a public clamour for bread by a hungry populace.[7] This and other similar incidents apart, it is hard to deny the special place royalty had carved for itself in the nation's public ceremonials. Indeed, the degree to which the persons and rituals of royalty were themselves a feature of English (primarily metropolitan) life can be seen in the ubiquity and popularity of royal themes in the broader field of popular culture – notwithstanding the low esteem of particular monarchs – or the low points in even Victoria's public standing. Travelling shows and fairs were rarely without their models, displays, tableaux and inconography of royalty.[8]

Then as now, London was quite different from the rest of the nation. So too were its public rituals and ceremonies. The world of the craftsmen, tradesmen and apprentices was not seriously displaced in these years by large-scale workplaces – and the rituals of the traditional workplace and practices continued throughout our period. Indeed, long after towns and cities in the north had begun to lose their older ceremonials through the pressure of urban and industrial change, the ceremonials of artisan life – the guilds, the fairs, feasts and holidays – continued in the capital. The rowdy fairs of Bartholomew and Greenwich, for instance, were not effectively outlawed until 1854 and 1857. Indeed in the century between 1750 and 1850 no fewer than sixty different fairs have been traced within a 15 mile radius of Charing Cross. Some, it is true were outlawed, but new ones grew and many were boosted by the coming of the railways and steamships. In 1843 for example some 200,000 attended the Stepney Fair. Yet what we need to note here is that these fairs were in origin, rhythms and style, closely related to the rural agricultural world. Here, in the greatest city of the land, patterns of rural ceremonials continued throughout our period, although admittedly under attack from men in local authority. Indeed it was the unchallenged antiquity of these metropolitan rituals which incensed some observers, as one said of Greenwich Fair: 'The whole mischief seems to have no other sanction but that of custom. . .'. This same fair was described (in 1825) as

> a never to be forgotten orgy of noise, swings, dancing-booths, oil lamps, fried fish, fat women, giants, dwarfs, gingerbread nuts, unappreciated actors, jugglers and acrobats, mud, dirt, drink, gin, beer, and skittles.[9]

A number of London fairs were heralded by major processions, many of them of great antiquity and yet none the less surviving as popular rituals until the mid nineteenth century (and beyond). And the noise and boisterousness of such processions was often accompanied by 'rough music' of the type so commonplace throughout pre-industrial European charivari. The temptation to offer simple generalizations about the destruction of traditional rituals in the modernizing city needs to be tempered by the fact that 'Charivaris were still a feature of the streets of England's largest city in the 1820's'.[10]

It is, of course, a legitimate objection that the public rituals of London were perhaps not typical of elsewhere. But, there again, the rituals of urban life (and indeed of rural life) were overwhelmingly local – often to an exceptional degree. For all the general patterns and national characteristics discernible to the modern historian's eye, the peculiarities and distinctions of popular urban ceremonials are perhaps even more striking. And yet it is further evidence of the continuing attachment to public ceremonials that, even in rapidly expanding industrial towns, customary rituals remained important and valued highlights on the local plebeian calendar.

Few towns could provide a starker contrast to London than Oldham. In the sixty years to 1851 its population had grown from 8000 to 51,818 – largely on the back of the local textile explosion. Renown as 'a place famous throughout all that country for daring and desperate wickedness', its people maintained throughout these years a firm commitment to their own brand of local radical politics – and to a calendar of traditional collective rituals, more especially the local Wakes. At the centrepiece of the summer Wakes were the ceremonies of rush bearing which, driven out of the churches and chapels by the emergence of an evangelical spirit hostile to such customs, became more secular. The parades of the local rushcarts, with the accompanying youthful boisterousness and drunkenness, survived throughout these years despite propertied hostility, a transforming local economy and an extraordinary boom in local population. Although it has been widely claimed that the coming of the railways effectively killed off the rituals of the Wakes, it now seems clear that, on the contrary, these rituals accommodated themselves to the new opportunities afforded by mass transportation. Oldham had a railway link by 1848; thereafter the Wakes festivities were transmuted into an epic trek to the seashore (until the 1870s, mainly for day-trips). And yet, although they were weakened, local Wakes ceremonials continued in Oldham itself. With the rising material prosperity of the last quarter of the nineteenth century local rituals were enhanced by further and more extensive travel to nearby pleasure spots, family and communal reunions and a host of proliferating commerical pleasures and entertainments. For our purpose, however, by the end of this period (and later) it seems pefectly clear that the rituals of Oldham Wakes constituted a 'rich and diverse festival where traditional and modern elements flourished together'.[11] In Oldham (and in other industrial towns in the Lancashire and Yorkshire textile regions) the rapid urban and industrial changes had not so much destroyed traditional communal rituals as incorporated them within the new system.[12]

Understandably, however, this did involve some major transformations, none more sudden or effective than the coming of the local police force. We have aready seen how, in the question of law and order, the new urban police forces proved highly effective in securing a more peaceable urban habitat. Precisely the same was true of public ceremonies. In Oldham in 1849 a borough police force of twelve men was formed. Thereafter the Wakes processions of rushcarts were accompanied by the new local policemen. Local processions henceforth (and almost immediately) became peaceable and 'respectable'. They remained faithful to the traditional rituals but were securely under the supervision and control of the new agents of law and order.

Oldham provides just one illustration of a national phenomenon. By the mid nineteenth century urban mass rituals had been 'tamed'. Indeed, it is striking how closely they came to resemble the civic and corporate displays. Local corporate organizations themselves joined the processions; local working men were now accompanied by the fire brigade and police force in colourful but well-ordered marching displays of local institutions. When these rituals were, in places, reduced to insignificance it was often because they were replaced in local public esteem by other attractions, especially by an extended holiday at the seaside. By 1900 an Oldham travel agent, extolling the virtues of Wakes weeks travel, optimistically advertised; 'Peking in 14 days, Tokio in 17 days; passengers leave Oldham (Clegg St) 2.42 p.m.' They went instead to Blackpool,[13] as indeed they had since the coming of the railways in the 1840s. These new collective rituals were to survive a century and more (and provided the author with his own annual childhood rituals and family enjoyments) until destroyed by the collapse of the local textile trades and the spread of car-ownership in the 1960s.

What enhanced the urban rituals of working-class life in the early nineteenth century was not so much those instruments of 'rational recreation' (so beloved of their social betters) but various developments of plebeian life which proliferated and were capable of transmutation from their pre-modern forms. This was true of the London fairs and of the

Oldham Wakes. It was also true of musical life. By mid century there were large numbers of popular urban ceremonials that were accompanied by music. Military music had, of course, long been a feature of civic and political ceremonials. By the mid nineteenth century the first flowering of modern working-class musicality provided a musical accompaniment to most public rituals, be it a Wake, or even local funerals. In Yorkshire, Lancashire, the north-east, the Potteries and south Wales in particular, there rapidly developed in the early nineteenth century the basis of that mass musical and choral performance (and appreciation) which survives to this day. It is true that this urban (and industrial) and, in large part non-conformist, plebeian musicality did not fully flower until the second half of the nineteenth century, but its rituals were clearly in existence in the early century. Indeed, many of the rituals of working-class life, politics, and processions of all kinds, took place to the sound of music. Music was, for instance, ideally suited to attracting public interest whether to the hustings, a travelling show or even to a more sedate procession (a tactic to be used later in the century by new religious organizations – the Salvation Army for instance – anxious to lure poorer spectators to their meetings). Indeed, the new brass bands (ideally suited to playing in public – in all weathers) were disliked by many, largely because of the associations with an older less 'civilized' series of popular rituals.

> I look upon all persons who admire brass bands [wrote one man in 1868] as possessing a primitive taste. Such music is only suited to precede a mob of inebriated rioters, revolutionists, electioneerists, trade unionists; and I hope the day is not far distant when it will cease to exist and be looked upon as a barbarism of the past.[14]

Once more, this ritual, a 'barbarism' of the past, did not go away but survived in an altered but strengthened form to become one of the main characteristics of late century urban plebeian life. And its importance was reinforced by the way it came to accompany a host of other plebeian rituals.

As ever more working-class institutions developed, each evolved its own distinctive ceremonials, in private and public, which in their turn became notable features of the urban calendar. Friendly societies and burial clubs – again with unmistakable eighteenth-century roots – flourished in the first half of the nineteenth century providing millions of working people with minimal protection against the ravages of illnesses and death. These organizations also evolved their own rituals, for instance to celebrate the passing of a member. Indeed, the working-class funeral offered a focus for a host of plebeian ceremonies that evolved from working-class organizations. Burial clubs, chapels, Sunday schools, ragged schools, trade unions, choirs and bands, all turned out in force, often in uniforms, in procession, with flags or instruments, in orderly and regulated homage to a dead friend or colleague. Such patterns of public, group mourning, traditional among the London trades, were now commonplace among the new industrial working class.

To a degree the ceremonials of bereavement were ubiquitous. All shades of society, in England (and elsewhere) adhered to the finely nuanced gradations of their own patterns of bereavement and interment. Yet it is tempting in considering the English rituals of death in these years to place too great an emphasis upon the protracted and complex ceremonials of royalty and the Court (with all their imitative influence lower down the social scale). Humbler people, unable to adopt the elaborate and costly dress and time-consuming mourning calendar of their social superiors, none the less celebrated death and bereavement through their own highly ritualized codes. Furthermore, the emergence of ever more working-class institutions, by the end of our period, provided a complex web for the development of new plebeian mourning rituals. What made these rituals more extensive and more public was a simple fact of urban geography; the new cemeteries of the second half of the century were established some distance from the centres of population and it was therefore necessary to *process* for some distance to the burial place.[15]

As with other areas of English public ceremonials, royalty and aristocracy displayed the most elaborate celebrations of death. Of course, in a very large measure their relationship with each other, but more especially with the public at large, was determined and directed by a complexity of public rituals and ceremonials. In the eyes of curious but always distant public, royalty (and the traditional aristocracy in their own localities) governed and held sway through ritual as much as power, though in fact the one was a reflection of the other. Royalty and aristocracy (then and now) were perceived and defined by outsiders largely through the medium of the public rituals performed by monarchs and aristocrats. And this was no less true in their rituals of death. The rituals accompanying the burial of the nation's élite were state occasions. The pomp, expense, and popular appeal of the major Victorian funeral processions were no less awesome than royal marriages and other state occasions.

Whatever the social level, dying and death formed one of the crucial rites of passage that demanded elaborate personal, family and communal rituals. While this may seem obvious (because so universal and contemporary), the celebrations of death provide further evidence of the continuities in English rituals in these years of major change. Bereavement was commonly marked by customs of dressing in (and draping the home in) black – with certain exceptions where white was used. The widespread use of flowers was, again, a universal reminder of the freshness of life in the midst of death, in addition providing an aromatic antidote to the smells of death and decay. All social classes celebrated death with drink and food – often an essential provision for those mourners who had travelled long distances. While the prosperous classes could celebrate the achievements of their loved ones in that array of obituaries, notices, plaques, memorials and the like, among the humbler sort, a lifetime's virtues (and vices) belonged to an oral tradition which began in earnest in the night- and day-long vigils around the corpse.[16]

The rituals of bereavement were both private and public. Death itself took place in the home, in the heart of the family with the comfort and succour of friends and neighbours. But its last rites were public and demanded certain observances of the outside world. At this level, the emergence of an ever more urban world had, by the end of our period, begun to impose new rituals upon the ceremonies of interment. The urban burial grounds of the early nineteenth century were, as we have seen, a cause of major political and medical concern; at once a source of squeamish distress and a major health hazard. Burial facilities were hopelessly inadequate and consequently, from the 1830s, new urban cemeteries proliferated (sometimes commercial but also, increasingly, civic) to provide a decent resting place for the dead. In the thirty years before 1863 no fewer than 2928 new cemeteries were opened.[17] Primarily for reasons of cost, they were established outside the existing town boundaries, thus breaking the age-old physical links between the church and the burial place. Graves were henceforth distant from the place of worship; relations and friends now had to make a conscious pilgrimage of remembrance to the graveside to celebrate birthdays or anniversaries. Similarly, interment itself was preceded by a journey and a procession (itself an extra cost which working people in particular could ill-afford). Urban life, and death, thus shifted the location of one of society's major and recurring public ceremonials. Interment was removed from the immediate neighbourhood. As with travelling greater distances to work, the centre of social gravity was shifting away from the immediate community and neighbourhood towards a more distant location. Burying a loved one on the edge of town (however more civilized the new environment) was perhaps one of the major breaks with former urban rituals.

The new cemeteries themselves became a prominent aspect of the mid nineteenth-century town. Indeed, the Victorian cemeteries, now so often overrun or vandalized, provide some of the most vivid surviving reminders of Victorian urban society. While so much else of the Victorian landscape has fallen victim to modernization and urban renewal, the cemeteries remain, in all English Victorian cities, in their recognizably original form. Of course, like the cities themselves the burial grounds grew in size. But unlike those cities they were planned, rational, landscaped and subject from the start to minute regulation and control; they were conceived, devised and created by architects and landscape designers, and their architecture attracted some of the finest of contemporary sculptors and designers. Indeed, it is not too fanciful to claim that some of the earliest and best efforts to transform the nature and quality of the urban environment was directed not so much towards the living as the dead. By the end of our period English towns and cities had their own, new outposts in the form of landscaped and tranquil cemeteries that provided a rural and yet orderly peacefulness within a short distance of the turmoil and confusion of urban life. It is no accident, for instance, that Victorians took to visiting the cemeteries, much as they visited the parks, for a walk and recreation. This was a major transformation. In the 1820s and 1830s there were few growing towns without their horror stories about death and burials. By mid century this had effectively changed and the English people were able to develop more modern, more 'civilized' public rituals of death and bereavement; rituals immediately recognizable to the modern reader. In this respect at least, there is clear evidence of the more civilized nature of mid century urban life.

It would be misleading to suggest that the continuities in social customs are to be seen in their clearest form in the more obvious public or communal displays. Death, like many other crucial rites of passage, was celebrated more particularly within the bosom of the family. Consequently a study of those rituals belongs more appropriately to the history of the family itself. Birth, betrothal, marriage – and death – each had its own (often specifically local) customs and observances which had changed very little for many years. But even if we set aside family rituals, English social life throughout these years was speckled by a host of rituals and observances which had existed since time immemorial. Escapades by apprentices (subsumed in some areas into the 'Wakes' tradition), national holidays that demanded certain local games, feasts or processions, Shrove Tuesday football matches, Whitsuntide collections by local children, local dances and songs, parochial fights and games, days given over to plebeian license or women allowed to take the initiative in amorous advances, bonfires, fireworks and carnival days when the social world was turned upside down, all (and more besides) formed an unwritten but potent calendar of rituals that provided an important framework for non-working plebeian life.

When in 1848–9 there was published an antiquarian account of the *Popular Antiquities of Great Britain* (an undertaking begun in 1795), the results ran to three large volumes; more than 1500 pages describing the 'traditional' pastimes of the people.[18] The accuracy of many of those entries is undoubtedly dubious. But, flaws and inadequacies aside, it was apparent to a number of people writing towards the end of our period that many contemporary rituals had origins which 'may be said to lose themselves in the mists of antiquity'.[19] Indeed, there was a developing passion in the first half of the nineteenth century to discover and document a comprehensive range of popular customs and rituals (perhaps encouraged by the sense of rapid change taking place in society at large). When, in 1836, William Hone published his *Year Book of Daily Recreation and Information,* he devised the subtitle, *Everlasting Calendar of Popular Amusements, Sports, Pastimes, Ceremonies, Customs and Events*. . . .[20] Again, we can discount some of his more fanciful entries, but none the less need to be aware that there were contemporaries impressed by the durability, the persistence and continuity in popular rituals; the 'everlasting calendar'. It is true that there was a different tradition; a more pessimistic interpretation which saw urban and industrial life purging the English people of their traditional ceremonies and rituals. While it is tempting to offer a balanced view – to claim that both sides were correct – what is impressive is the degree to which current historical investigations are actively stressing the continuities rather than the breaks. Even by the end of our period the people of urban England seemed remarkably addicted to rituals that were recognizably those of their forebears and which formed a complex cultural inheritance from one generation to another.

11 *Family life*

In the mid eighteenth century the family unit was the pre-eminent social (and in large measure economic) unit of English society. It was, in the words of Roy Porter, 'the elemental unit of living and dying, reproduction and socialization, education and business, love and hate'.[1] This was no less true a century later. It is tempting to imagine that the family, like so many other human organizations, was utterly transformed by the seismic changes in economic and urban life, more particularly by the coming of industrialization. It is equally tempting – and no less misleading – to accept as widespread and typical the image of mid Victorian domestic harmony and bliss, so assiduously disseminated by the idealogues (male and female) of bourgeois life. None the less, the mid Victorian family reigned supreme – whatever its imperfections, tensions and shortcomings. It did not, however, please Karl Marx – despite the heroic efforts of his own family to keep him in the style to which he grew accustomed.

> On what foundation is the present family, the bourgeois, family based? On capital, on private gain. In its completely developed form this family exists only among the bourgeoisie. But this state of things finds its complement in the practical absence of family among the proletarians, and in public prostitution.[2]

Successive generations of investigators, taking their cue from this and other assorted denunciations of family life, have been at pains to denounce or dismiss the Victorian family.

We need not, however, accept in its entirety this bleak image of bourgeois and plebeian family – but it would be equally wrong not to examine the implications of those criticisms. The family, of all social classes, was clearly affected and modified by the major social and economic changes in society at large. Indeed these changes – whether in the form of wage levels, under- or unemployment, migration, hunger, working, or housing conditions, immiseration or rising prosperity – made themselves felt immediately and most directly on the structure and fabric of the family. It was the ease, or inability, of securing a decent material life for one's family which was, and remains, the crucial criterion for measuring levels of prosperity or austerity.

Long before our period began – and long after – the basic family unit, of the married couple living with their offspring, the nuclear family, was the basic norm of English social experience. Moreover, this has been a fundamental feature throughout the western world for at least three centuries.[3] However notable the changes in family life and structure in the years under study here, they form only minor alterations in the durable and resilient structure of the nuclear family. The English family – plebeian or bourgeois – was neither created nor destroyed by contemporary economic change in these years. It did, however, suitably accommodate itself to the changing demands made of it. And in 1851, as in the 1770s, it continued to be the pre-eminent crucible from which future generations emerged, equipped with their family inculcated values, skills, attitudes, duties and responsibilities – and doubtless neuroses – which enabled each individual member to survive in his or her own mean or prospering condition.

At all levels of society family life was shaped and constrained by the particularities of economic conditions. For the very great bulk of English people in these years, rural and urban life was harsh and unrelenting. Of necessity the plebeian family was carefully organized to extract the maximum physical or skilled returns from its various members – just as long as they had the strengths, in childhood or old age, to lend their efforts to the family's well-being. Much depended, of course, on the nature of the family's work – which might be varied and diverse – but a working man would, in general, need the hands of his wife and offspring. The fact that generations of tiny children were forced into arduous work, at home, in workshops, factories or fields, tells us nothing whatsoever about the emotional relations between parents and their children. Economic need, not cruelty or lack of affection, was the prime motive. Just as, since time immemorial, infants had worked within the family unit in rural activities, so too were the growing armies of urban children put to the varied tasks of assisting their families. And as the factories came to dot – though not yet to dominate – the skyline the young accompanied their mothers and fathers into industrial occupations. That mothers and fathers took their young into the factories, workshops, mines or whatever, reveals not insensitivity or harshness on their part, but the degree to which work and the need to survive – as a family unit – was all important and pervasive.

The English economy, however described in generalized and national terms, was the sum of an incalculable volume of family economic activity. Indeed the need to be able to maintain a family – or at least a spouse – was a prime consideration in the decision to marry. Artisans and working men delayed marriage until economically able to sustain their new responsibilities. Throughout the eighteenth century pre-marital sex seems to have been a temptation which was resisted by most and, despite the decade between puberty and marriage, the incidence of illegitimate births was relatively small.[4] Yet, as the century advanced bastardy rates increased, an indication perhaps not so much of a changing moral or emotional climate, but of rising levels of fertility, related to improvements in health and food, that were themselves responsible for the overall population growth.[5] The stigma of bastardy, for mother or child, cruel and painful as it now seems, was in effect the mirror-image of the contemporary importance and social significance widely attached to marriage and family life. The unmarried mother and her child were the cruelly-isolated reminders of a conflicting image that seemed to threaten the family-based structure of society. These unhappy women paid for society's hostility to their 'short comings' by enduring the harshest of personal conditions. It is true that the bastard offspring of polite Georgian society was, generally, materially well cared for and socially accepted.[6] Poor illegitimates and their mothers, on the other hand, were forced to endure the most abject of circumstances. When demographic data came to be collected in the course of the nineteenth century it became apparent that illegitimate children and their mothers suffered most acutely all the dangers and pains, of childbirth, infant sickness and mortality, and maternal mortality. It is true that the highest levels of recorded illegitimacy remained rural but, wherever it was found, the unfortunate women and their children could expect little medical or social help. It was a common rule, for instance, that unmarried mothers were to be denied the facilities of the charitable Victorian lying-in hospitals. Not surprisingly then, they left their mark more clearly than any other social group in the fearful details of infant and maternal death statistics.[7] The harsh treatment they were forced to endure provides, in its crudest form, not merely an indication of contemporary attitudes to sex and sin but also offers a clear reflection of the values attached to family life and, conversely, the determination to penalize those who dared to deviate from it.

There are a host of ways in which we can measure the contemporary attachment to the family. The family ideal became the crucial centrepiece of Victorian bourgeois mythology – recorded time and again by the scribes (many of them women) – of contemporary domesticity. It was, similarly, recorded throughout a vast range of late eighteenth- and nineteenth-century iconography. Indeed one of the most revealing of pictorial documentations of contemporary life was the family portrait – later photograph – which captured not only the physical appearance of the various members of the family but also their relative ranks and relationship one to another. While it may be true that the sombre, early photos of family groups in the 1840s and 1850s, gathered round a dominant and stern paternal presence, contrasts sharply with the more informal paintings of the late eighteenth-century, it may also tell us as much about the physical restraints and rigidities of early photography. Whatever interpretation we adopt, the central and irrefutable point remains: generations of Georgians and Victorians were captured for posterity as part of their family group. Portrait painters and early photographers sought to represent contemporary images through the family unit.

Such evidence is less easy to come by lower down the social scale. Even there, however, the evidence is overwhelmingly conclusive of the importance working people attached to the basic family unit. This was graphically illustrated after 1834. Among working people, one of the severest complaints against the new Poor Law was that it physically dismembered the pauper family by dividing up partners and sexes inside the bleak walls of the new 'bastilles'. The family also looms large in the recollections of nineteenth-century autobiographies. It is also possible to cull similar evidence from other unlikely but parallel issues. Among those increasing numbers of people who, from the 1780s, wished to free the slaves throughout British possessions, it was widely accepted that one of black slavery's most damaging vices was its alleged destruction of, or obstacle to, the slave family. We now know, however, that the nuclear family was firmly established in those slave communities by the late eighteenth century – as it was in the US South in the early nineteenth. None the less, the assumed absence of the slave family (a view which normally derived from the absence of a formal Christian marriage ceremony) was a bitter complaint in the abolitionist litany. In 1792 in a draft proposal for black freedom, submitted to William Pitt, Edmund Burke suggested that

> all negro men and women, above eighteen years of age for the man and sixteen for the woman, who have cohabited together for twelve months upwards, or shall cohabit for the same time, and have a child or children, shall be deemed to all intents and purposes to be married . . . .[8]

Similar attitudes were to be found throughout the nineteenth century whenever that expanding army of missionaries, facing 'primitive' peoples, thought they detected few traces of a conventional British marriage or family structure. Child marriage, arranged marriages, polygamy (and more) offended the British Christian imperial instinct and became part of the missionary impulse to convert the 'pagan' world, to the benefits of proven British practice – none more proven and valued than monogomy and the family unit. The world, it was hoped, would turn a British imperial red and in this process of 'civilizing', the world's indigenous peoples would be recast and remoulded through the crucible of the British family.[9]

There exists an abundance of similar evidence and it is possible simply to accumulate more and more contemporary impressionistic evidence to illustrate the firm attachment of contemporaries, of all social classes, to the existence and enhancement of the nuclear family. What needs to be said, however, is that, in the last resort, it is the demographic evidence about structure and size that offers the clearest case for the durability of the English nuclear family.

Recent work by Wrigley and Schofield shows that one of the key determinants behind population growth (in this period and others) was the high levels of 'nuptiality' – of people marrying. From the mid seventeenth century onwards fewer people remained unmarried, and the increasing incidence of marriage had an obvious but major impact on reproduction (in the absence of any significant attempts to control fertility).[10] But since marriage was normally entered into only when economic circumstances allowed, this assumes (with variations in times and place) an upturn in economic fortunes. Thus we need to place the history of the family firmly within the context of contemporary economic change.

What makes a study of the western family doubly difficult is the sheer burden of mythology and propaganda – both historical and present day – which has ensnared the family in a cocoon of unreality. An idealization of family life was actively promoted by apologists for nineteenth-century domesticity, while more recently the modern media have for various, though largely commercial, reasons purveyed a similar image. Much of that imagery is often pure caricature. And yet, like all caricature, to be effective it needs to have a kernel of recognizable reality. Take for instance the matter of male dominance.

In many respects the family was a male-dominated institution, but such a judgement can mask the crucial roles traditionally played by women and girls, especially in certain regions and industries. We have already seen how the economic role of many working-class women was an integral feature of the overall family economy. After all this had traditionally been the case in the pre-modern world. Similarly it remained the case in agricultural life into the twentieth century. Certain key industries –

textiles for instance – were notable employers of female (and young) labour, with the consequent importance of female earnings within local family life. Of course, the strictly economic value of female labour can only partially be measured in terms of work and income outside the family. Perhaps the most important economic role played by women was the management of domestic funds and labour. At this level women of all social classes shared a common role, however contrasting their economic fortunes. Indeed perhaps the clearest indication of the centrality of women within the family economy and management is to be seen in the haste with which widowers, particularly working men with children, remarried. For a working man with children (and a new arrival was all too often the cause and occasion for the loss of a wife) it was vital to secure another spouse to take over the indispensible role of cook, child-minder, manager – and wage earner.

The middle- and upper-class woman occupied no less important a position, despite the fact that she, unlike the working-class woman, did not undertake paid work outside the home. The more prosperous and bigger the household, the greater and more complex the demands of management and financial regulation. Domestic management, minute regulation of servants (some of them of course men) supervisions of child-rearing – much of it actually done by nurses and nannies – book-keeping, victualling, etc., all demanded of the woman of the prosperous house a skilled and time-consuming daily routine. It is, of course, easy to dismiss such women as merely enjoying the unearned fruits of their husbands' labours and the efforts of their domestics – but this would be to misunderstand the dynamics of family activities. Nor is this to deny the myriad of cases where women found the demands of domestic management stifling and pointless. For such women the outlets were few and far between: church, chapel, Sunday schools and an expanding range of charitable 'good works' that soaked up the untapped energies and frustrated ambitions of generations of English propertied women. For some, however, all this was no real outlet for their interests and ambitions. In the words of one of their more famous members, Beatrice Webb,

> According to the current code, the entire time and energy of an unmarried daughter – especially if she was the responsible mistress of the home – was assumed to be spent, either in serving the family group, or in entertaining and being entertained by the social circle to which she belonged.[11]

This formidable young woman broke with this mould by becoming, in the first instance, a rent collector in the East End.

Manuals and handbooks for girls and women of the propertied class took great pains to explain and amplify the role and obligations their readers would be expected to fulfil within their families (families of their parents, and later their own). Of course, the prime lessons of female conduct within the family were, like most other family attributes, simply acquired by growing up within the family unit – though, again, learned most directly from the mother. The family was as much an instrument of education (in the wider sense of socialization) as it was a forum for integrated economic activity or emotional development. None the less, the informal lessons of family life were reinforced by girls' books and manuals. From child-rearing to jam-preserving, from obedience to God/parents/husband, to nursing, the literate girl was provided with detailed guidance and advice.[12]

> It is in acquiring a knowledge of household affairs, chiefly, that your body is to be educated. Young girls often have wrong notions about this matter, looking upon *house and work* as mere drudgery, only fit for servants.

As if to convince the young female reader of the value of such lessons, it was claimed

> The truly useful woman has an influence upon those about her, and indirectly on society. In every sphere of life woman is eminently useful and eminently important.[13]

But these general virtues of the woman's role were to be honed and shaped primarily within (and for) the immediate family. Thus the family was seen as the most formative element in shaping qualities necessary for the well-being of society at large. The family was, in a sense, a miniature of wider society – ideally if not always in fact.

Within this idealized imagery of Victorian bourgeois domesticity, the family was able to pass on its lessons and its manifest virtues to those lower down the social scale. It was assumed that

> Each household needs a different ordering, and each class of society has its own requirements; but it may be said that every class teaches the one immediately below it. . . .[14]

In this manner servants would acquire the application, skills and habits to manage, in their time, their own humbler households. Working-class girls were the objects of endless moralizing and instruction, both from their immediate employers (if, as so many were, in domestic service) and from whatever private or charitable agencies they encountered. Even the workhouses, the instruments of so much damage to the poorest of families, sometimes sought to inculcate the skills necessary to family life: knitting, sewing and household work. Similarly, so did the charitable schools (and of course at a much later date, the compulsory Board Schools). Indeed one of the most persistent and telling criticisms of early nineteenth-century factory life was that, by driving young girls from their homes, the factory denied them training in the necessary family-based skills. Writing of the Potteries, it was claimed,

> The girls are prevented, by their removal from the home and from the day-schools, to be employed in labour, from learning needle work, and from acquiring those habits of cleanliness, neatness and order, without which they cannot, when they grow up to womanhood, and have charge of families of their own, economize their husbands' earnings, or give to their homes any degree of comfort.[15]

In charity and industry schools throughout the eighteenth century, sewing had been an important part of the girls' education. This continued to be so in the National and British Society's schools (founded in 1811 and 1814). Even as late as 1862, in the Education Department's Code for elementary schools, needlework was obligatory for girls.[16]

Throughout the nineteenth century, in voluntary and later compulsory schools the guiding inspiration behind the education of young girls was to fit them for their role within the family life. Of course, in stark utilitarian terms, however inadequate it may seem in retrospect, this ideology of female education made functional sense. By the lights of contemporary values and needs, it seemed both obvious and necessary to equip girls, of all social classes, with the social and economic skills they would require as adult (or adolescent) domestics, wives or mothers. Yet the relevant point for our interests is that the ultimate inspiration behind girls' education, formal and informal, was the family. Whatever value such training might have in society at large was incidental – however welcome. Its prime purpose was to contribute towards the continuing well-being and benefit of the family, *tout court*.

However important the woman or girl within a family's economy and management, it was a role strictly subservient to the dominant presence of the head of the family – the man. To say that man ruled – pre-eminent if not supreme – is merely to repeat the obvious. Within the family – as in society at large – the dominance of the adult male was undisputed and was shaped by the complexity of the law, confirmed by religious observances and secured by economic dominance.

> Notwithstanding frequent absence, my father was the central figure of the family life – the light and warmth of the home. How well I remember how we girls raced to the front door when we heard the wheels on the carriage drive.

This description, by Beatrice Webb, may have had the happy glow of hindsight but in essence it captured the reality of male dominance (even in her case, in a family of ten women – mother and nine daughters). It was an assumption of family life that the male was (or ought to be) the head, the breadwinner (even when he was not), the source of discipline and control (even when in practice it fell to the woman), and the general protector (even though he might be incapable). The corollary of male dominance was female subservience and obedience (notwithstanding their particular social or economic contribution). Writing of his brother in 1794 a York man noted,

> His first duties are to behave well in the station in which divine providence has placed him by acting as a good Husband, a good father and a diligent and upright man.[17]

This was, in unmistakable outline, a clear assertion of the contemporary view of patriachy.

Perhaps the most important manifestations of male dominance took the form of the legal relegation of the woman to a greatly inferior role within the family. Blackstone's famous dictum was widely accepted; 'that the husband and wife are one person and that person is the husband'. Married women had no legal rights to sue or be sued or to make contracts, their property automatically passed to their husbands unless agreements specified to the contrary. It is well known that women of property (unmarried or widowed) often brought essential capital to a new marriage and its attendant business interests. But female independent access to, and control over, money that was specifically their own was often curtailed or controlled (via trustees for instance).[18] Needless to say, such problems did not really affect the lower orders of society in any significant way. Indeed the distance and remoteness of working-class life from the process and protection of the law, however discriminatory against the woman, was a crucial factor in exposing innocent and hapless working people to the dangers of financial tricksters and fraudulent schemes.

There was then a paradoxical contrast in the role of women within the family – and one which was becoming sharper with the emergence of new economic structures and processes. The woman was the pillar upon and around which family life was based; a role praised and lauded throughout these years. It was not however an equal role. Time and again, in manuals on the management of the middle-class family, girls and young women were given copious detail on how to train themselves and how to undertake their role as managers and controllers of family affairs. And yet above them, in any uncertainty or dispute, hovered their closest male – to be relied on and consulted when they themselves failed. Women were told for instance 'to appeal to your father or brother should you be unable to secure obedience from the servants'.[19] The nearest male – husband, father, brother – was to be accepted as the most significant person. The practice, of course, may have been quite different; with women turning to and relying upon other women for succour and advice. This was especially true within working-class homes where the network of female ties – of kinship, neighbours and friends – provided that web of assistance that formed a body of advice in times of need and a proven body of expertise and reassurance to be acquired in an informal though none the less seminal fashion. Whatever flaws existed in a particular family – flaws often brought about by the unpredictable blows of fate – they could in some degree be offset by help from the wider neighbourly community – and the support of other women. In this way working-class women came to the aid of other local women in times of crisis and need – times all the more common then – of shortages, childbirth, illness and death. Indeed the rituals that accompanied these distinctive rights of passage within family life were above all else, conducted and shaped by the women of the house and neighbourhood. In a society which lacked those formal state or private institutions which, today in the west, attempt to cope with or cushion such problems, the family – in its immediate setting of kinship and neighbourhood – was the prime instrument of medicine and social care. And within that setting women of the family ruled supreme.[20]

It was the women who cared for the sick – much more of a persistent and lingering problem then than now. It was women who looked after pregnant women, in childbirth and who nurtured mother and child afterwards. And it was women who organized and dealt with the highly complex rituals of death and bereavement. Even when, in all these situations, a family was lucky or prosperous enough to have recourse to professional help (and as we have seen these years witnessed the professionalization of many of these functions) the professional assistance, doctors, midwives, nurses, undertakers, dealt with the women of the family. Indeed the crises of family life, like its daily and ritual economy, were primarily the preserve of the women of the family and neighbourhood. In poorer homes these same women 'doubled' as semi-professionals, their simple experience standing them in good stead in their capacity as midwife, nurse or layer-out of the dead.[21]

All of these functions, overlapping and complicated, were in addition to the other prime role of women within the family economy, either as wage-earner, cook, child-minder (or more likely a combination of all three). Yet for all that there was no suggestion, in law or social acceptability (at any social level), that woman's role was any other than inferior to man's. The parallels between woman's lot and bondage are beguiling (though ultimately misleading), for theirs was an economic and social role that was major and ubiquitous, while having a legal *persona* which was insignificant, secondary and in some cases non-existent.

Foremost among the functions of the family was that of child-rearing – of socializing the young to their particular social and sex role. It was within the family that children acquired their initial and normally abiding lessons, skills and disciplines, useful for themselves, the family unit and society at large. But it was some measure of the transformations in society that, increasingly, towards the end of our period emergent state and private endeavours began to establish institutions to offer the skills that many families could not or would not pass on to their children. This was obviously the case with the growing formal insistence on literacy and with certain of the newer skills in industrial and technical operations. Sunday schools, charity schools, ragged schools (and later the Board schools) the reformed public schools, and many more sought to inculcate the social skills required by both sexes and all social classes. The reform of educational institutions, from the civilizing of the public schools through to the massive proliferation of working-class Sunday schools, can be seen at one level as a reflection on the inability of the family fully to provide their offspring with all the social/educational qualities they would need in the adult world. In this sense, formal education began to provide a supplement to the family. For the prospering middle classes, tutors and private academies, later the public schools (much later still for their girls), for working-class boys and girls, charitable, later 'state' schools, began to reinforce the broader education of family life by providing a learning which encouraged more varied skills.

Traditionally the family had provided basic training and education, however broadly defined. The mother had been crucial in the provision of basic literacy. Even in humbler homes, books and book learning were more common than might be supposed. As David Vincent has shown 'Few homes of the labouring poor in pre-industrial Britain were devoid of literature.'[22] The Bible, Prayer Book, Hymn Book, and other books of a religious nature were commonly found in working-class homes. Ownership of books did (and does) not of course offer an accurate guide to readership. But there is other firm evidence of high levels of literacy, although always with regional and sexual variations around the norm – and this evidence can be traced back to the seventeenth century.[23] But as industrial change advanced, book learning and the possession of a small library became one of *the* most prized of skills and possessions to be handed down within the family as a special item of family ownership and distinction. None the less, it would be wrong to deny that this was a distinct *minority* tradition within working-class life; the better-off and more skilled the working-class family, the more likely they were to be literate. The development of the factory system made it ever more difficult to integrate work and reading on the pre-industrial model.[24] And, of course, this same factory system divorced work from the home, and it was thus important to offer a new educational diet that would provide intellectual nourishment for new generations of children. Thus there developed the need for factory schools – and more. Such factors applied manifestly only to working-class children, although even their social betters increasingly felt the need to provide their offspring with a fuller and more rounded education to equip them for a changing world.

Formal education, however, was only part of the wider socialization of the young. Perhaps the most important role was the acquiring of economic skills. In the case of domestic skills, these were passed by mothers to daughters. In more prosperous homes the skills acquired by girls of the house were those of household management and control over labour. But there were also the practical lessons for girls of all classes – thought to be essential for any family life – cooking, needlework, and 'domestic science'. Naturally enough lower down the social scale this tended to be much more rudimentary. Often too educational opinion felt the need to encourage such skills among working-class girls – in the workhouse or school – from a fear that the working-class home was providing an inadequate training for domestic harmony and security. Indeed one impulse behind the ever more insistent call for formal education was to make good the disadvantages inherited by poorer children who were ill-equipped to enter the adult world. Boys were similarly encouraged to acquire those economic skills – of a plebeian or middle-class nature – that they would need as breadwinners. This was especially striking in artisan circles and family businesses where boys would be apprenticed to a skilled or trading father, before themselves taking over their father's role. This had, after all, long been the customary pattern in the inheritance and continuance of artisanal skills. Wherever such skills survived – even in a new industrial town like Preston – they were often part of family activity. We ought not to imagine, however, that work within a family network necessarily safeguarded a child from the harshest of conditions. Such children,

> brought up to and put to a trade, often for long hours, from an early age, they had no other skills and had nowhere else to turn for equally remunerative employment.

In addition, hours and working conditions were often appalling.[25] This was, clearly, an acute problem in decaying artisanal or domestic industries. Child labour continued in its most exploitative and rapacious form, not so much in factories – where in the course of the nineteenth century union and Parliamentary control managed to check its worst excesses – but in those surviving occupations and sweatshops which thrived on cheap female and child labour. A great deal of this labour was undertaken by the family unit.[26]

It is easy to see in such bleak evidence proof of the indifference, hostility or lack of affection by working-class parents to their young; a view which unites certain modern historians with a number of early Victorian commentators. Moreover, it is a view which might at first glance appear to gain support from the family-based nature of early factory life. Time and again, parents took their sons into their workplace. In the textile trades, heavily dependent on young and female labour, the family group was widely represented. For many working-class families there were considerable economic opportunities afforded by working together within the same workplace.[27] Paradoxically, however, it was also true that in time of buoyancy, factory life (again, notably in textiles) could provide young adults with sufficiently attractive earnings to enable them to break away from their parents.

Recent research has added detail to a point commonly echoed by contemporaries in the 1830s and 1840s.

> Children frequently leave their parents at a very early age in the manufacturing districts. Girls of sixteen, and lads of the same age, found that they can enjoy greater liberty and if not greater comforts, that at least they can have their own way more completely in a separate home, and these partings cause little surprise or disturbance.

Changes such as these had ramifications on the age of marriage of the departing offspring, yet for all the contemporary and subsequent commentary that such transformations had reduced the family to little more than a joint-stock company, the family remained healthy, durable and resilient.[28]

The early departure of the young to industrial work, in conjunction with other evidence, has often been interpreted as a sign of changing emotional ties within the family. There is, however, an abundance of contrary evidence; of deep and abiding love between parents, children – and then their own offspring. Naturally enough it is extremely difficult to measure this in any meaningful way, but time and again the evidence demands our attention. There was, for instance, communal dislike and opposition to those who did not treat their children or parents in an acceptable and affectionate manner.[29] Early nineteenth-century commentators provide a great deal of testimony about the love of parents for children – even in the most distressing of conditions. Writing of Liverpool, one man noted,

> It is really only the professional beggars among the Irish who will use their children; the respectable Irish are very tender to them, and will try to get them into schools, or to bring them on in the world, and do as much for them, as the parents of any other country.

Similar evidence was forthcoming about love between parents and children in the industrial north. 'Nothing can be more warm and keen than the affection of parents throughout the cotton districts for children, so long as they continue children. . .'.[30] It becomes perfectly clear that the bitter working-class resistance to the worst factory conditions for children in the 1830s and 1840s flowed in the first instance from family affection and also from the more generalized feeling that such outrages were wrong in themselves and quite inappropriate for the young. Time and again parents sought to defend their own children within the factories and, when necessary, they forcefully restrained and even attacked any managerial wrong-doer. Working people resented conditions endured by their children, but their options were of course strictly limited: 'but I was a poor man, and could not afford to take him away, having a large family, six children under my care'.[31] Working-class parents were torn hither and thither: between their love for their offspring and the natural desire to protect them, and the countervailing need to secure a decent life by harnessing everyone in the family to gainful employment.

The intensity of love within the family can also be gauged from an unlikely source. In that most acute of personal and family crisis, death, we find abundant evidence of the deep emotional ties within the family. It is seductive to see in the plebeian return to work during and immediately after bereavement, signs of a less than emotional commitment to the deceased. Of course, grief and mourning were severely constrained by the prevailing economic context; by the need to continue daily life. It is true, that,

> the poorer the family, the greater is the strain on the sensibilities of its members, but that is not the same as arguing that poverty obliterates affection.[32]

Indeed evidence from working-class autobiographies suggests that the harshest of material circumstances could, to the contrary, serve to intensify rather than blunt, emotional feelings. Sometimes, however, working-class writers had trouble expressing their emotions adequately, though often Biblical imagery provided a perfect and ready-made vehicle for such accounts. Yet even the barely literate could sometimes capture emotion in a unique fashion:

> I was as fond of my wife Has a
> Cat is of New Milk.[33]

Quite apart from the technical difficulty of expressing emotional responses, there was a sense in which working-class writers simply regarded such matters of no interest to outside readers. Others simply asserted their love, more especially the act of falling in love. 'I was smitten at once. It was love at first sight.' We are also told of autobiographers who 'fell deeply in love'. John Clare was 'head over heels in love'. Another wrote of 'my new passion of love', and yet another recalled 'the Joy and Bliss of loving'.[34] This is not to claim, however, that such romantic love was the sole and determining factor in persuading men (and the autobiographers were overwhelmingly male) to choose a spouse; the decision after all involved two partners. And prime importance was normally attached to economic considerations.[35] None the less, however blunted by subsequent material hardship, romantic love was indisputably a motive in the selection of a marriage partner.

The subsequent family group, however beset by troubles and hardships, also saw the flowering of new emotional attachments – to offspring. No amount of special pleading can disguise the contrary and overwhelming evidence of deep and abiding love for children. This was strikingly so in the case of the death of children. The radical essayist Leigh Hunt wrote;

> There are griefs so gentle in their very nature that it would be worse than false heroism to refuse them a tear. Of this kind are the deaths of infants.[36]

Infant and child mortality was such that it taxed the emotional lives of all families to the limit. And yet, perverse as it may seem, the death of children provides further detail, if more is needed, of the role and durability of love within the family. Time and again families of all social classes were cast down by the lingering, sudden or anticipated death of their babies and children. For the poor, the brevity of mourning cannot hide the anguish and grief, however pressing the need to continue with daily life.

The death of a child was, however, only the most frequent of family bereavements. In a society where diseases and illnesses – and of course childbirth – swept away people who would, today in the west, scarcely be troubled by the ailment, death struck often, and sometimes savagely. Samuel Bamford, at the age of 7 lost a brother, sister, grandfather and uncle in an epidemic.[37] In one sense it may seem pointless (because so obvious) to reiterate the emotional ravages of death within the family. But it needs to be asserted because this grief is the reverse of affection; a manifestation of despair for the loss of a loved one. Here again we are faced with a mountain of evidence confirming the primacy of love within the family – in all social classes. Love frustrated, love rebuffed by illness and death – these and more are liberally documented not merely in early Victorian literature and iconography but also in what might at first sight appear to be unyielding sources about working-class life.

Death and affection bring us once more to the harsh economic realities that formed the determining barriers of contemporary family life. Death – and the illnesses preceding it – could plunge a family into economic distress. This was obviously most acute within working-class families, but it was also true in the lower reaches of the middle class. A dead wife created an emotional disaster – and the loss of a child-minder, housekeeper and provider. A growing child, now dead, removed potential income and help. The loss of a father and husband could throw the survivors into the maws of the local workhouse. David Vincent has argued

> The loss of a close relative was so bound up with the material problems of life that at worst it seemed no more than an intensification of the miseries of existence.[38]

For untold legions, the ramifications of illness and death formed a different, perhaps more intense, variant of the harrowing difficulties of survival itself. But to suggest that this was conducive to the deadening of the emotions, a denial of love and affection within a particular family, is to ignore other revealing testimony. Men and women grieved (often silently) the losses within their families; mourning them in proportion and relation to their love and affection. Death was (and of course is) the catalyst for those subterranean emotions which rarely break through the surface.

A further indication of the importance of family life can be gauged from the rituals of the family cycle which were important not only within family life but also within society at large. Indeed, in many respects family life was shaped, guided and even defined by rituals – many of which are obvious because they remain familiar to any modern reader. There were well-established rites of passage which heralded the origin and founding of a new family group; new births were similarly celebrated as were the departure of young adults to form their own families. And, of course, death was itself characterized by highly elaborate rituals of mourning and bereavement.

Each of these family rituals was ancient; patterns of human association and relationship which, whatever their changing detail, reveal through their very durability the remarkable continuities in English life across many centuries. Many family rituals undoubtedly changed in these years, but what is perhaps even more striking is the degree to which many more persisted in their recognizably traditional and unmodernized form. And this was to remain fundamentally true until the end of the nineteenth century. Yet we need also to remind ourselves that the rituals of the human life-cycle – betrothal, marriage, birth, christening, bereavement and burial – had distinctively local (often parochial) forms, making it difficult to sketch a generalization to cover the whole nation.

Betrothal had long been marked by the giving of a ring or the exchange of valuables; those betrothed often distinguished themselves from others by wearing a particular item – or flower. Marriage, too, had, since time immemorial, been characterized by the giving of a ring and the eating of a wedding cake. So too were the employment of bridesmaids – and men for the bridegroom – and a profuse use of flowers during and after the ceremony. Even when describing 'Popular Antiquities' in the mid nineteenth century, English antiquarians were able to document an amazing variety of traditional marriage rituals that continued to thrive throughout England. These included the use of garlands, the giving of gloves and, more saucily, garters.[39] Many such rituals were, of course, religiously sanctioned, but it is an indication of the attachment to ritual as a communal sanction at the founding of family life that even those entering informal liaisons (rather than formal marriages) had their own distinctive and binding 'marriage' ceremonies.

The birth of a new child and its christening was also marked by family-based festivities although, again, such rituals were naturally limited by material conditions. Food and drink were normally provided.

> At the christening entertainments of many of the poorer sort of people in the north of England (who are so unfortunate as to provide more mouths than they can with convenience find meat for), great collections are often-times made by the guests, as such as will far more than defray the expences of the feast of which they have been partaking.[40]

But it was in death that we find some of the more sharply defined family (and community) rituals. Death was a much more common experience within most families than in the west today: babies, mothers in childbirth, young children, apparently healthy young adults, in addition to the old were commonly swept away; a process which had to be accommodated, rationalized and, above all, handled in a fashion that was both practical and yet tolerant of prevailing social *mores*. Throughout the century the determination to secure a 'decent' funeral for one's family members was a ubiquitous element in English society – even among the poor who, to the unsympathetic outsider, ought to concentrate on more utilitarian concerns. Few were more critical of contemporary plebeian celebrations of death than Sir Edwin Chadwick, the inspiration behind the 1834 Poor Law Amendment Act. This act blighted the family life of the very poor, nowhere more cruelly than in denying to them the time-honoured rituals of family bereavement and interment. Pauper families were separated, in death and in mourning, and the dead consigned to an unmarked and communal grave. The pauper's death and grave thus transgressed some of the most highly-prized values of plebeian life; allowing no place, occasion or time for family (and communal) mourning practices and for those rituals of remembrance (on birthdays, anniversaries and the like) that were so important throughout society.[41]

Outside the workhouse, what underlined the significance of death as a family ritual was not merely the pain and grief at the loss of a loved one but the simple (and often misunderstood) fact that death overwhelmingly took place within the home. Indeed it was thought appropriate that death *should* be at home – yet another reason for the hatred of the workhouse. After all there were few institutions – hospitals or the like – where the ailing or aged could secure treatment, cure or find a peaceful resting place. Thus the family home was the place for caring for the sick in their last days and hours, for laying out the dead and observing the local rituals of bereavement – just as it was the place for childbirth and child-rearing.

English people died (accident apart) in the bosom of their families – as the iconography of death clearly shows. Local, social or class customs decreed that there should follow the elaborate family rituals, more extravagantly observed higher up the social scale and reaching its lavish apogee in and around the royal court and aristocracy. And yet, however humble and poor the home, the rituals associated with death were none the less elaborate, socially binding and relatively costly, though rarely as public or as expensive as among their social superiors. But whatever class we examine, it is clear that the observances of death and mourning were most significant as a *family* ritual. And this was perfectly illustrated by the sharp gradations of mourning expected of each particular mourner, all depending on the proximity to or distance from the deceased.

There were, in addition, a host of other rituals and ceremonies which had such enduring strength and longevity because they were ultimately rooted in the family. Indeed, although many were viewed as national holidays (Christmas and Easter for instance) they were in fact primarily family celebrations. One commentator noted in 1800 that of all feasts and amusements

> the principal are at Christmas, when the greatest hospitality prevails among the villages; every family is provided with goose pies, mince pies, and ale.

It was, as it remains, a time for families to gather. Significantly servants were allowed home to visit their families. And it was the time for families to join with each other in celebration. Writing of Christmas in Northumberland, one early nineteenth-century observer remarked that it 'consists of feasting and social intercourse between neighbouring families'.[42] It was undoubtedly true that the nature and attitudes towards Christmas changed in the 1830s and 1840s, becoming more child-centred, more sentimental and more burdened with recognizably modern artefacts (Christmas trees, cards and the like). But this was only a change of degree. Furthermore, these changes merely reinforced the commitment to Christmas as a family affair.

> Perhaps the greatest characteristic of Christmas Day [wrote one man in 1868] is the very general custom of regarding it as a domestic and family festival. The thoughts of men seem to turn on that day more especially towards home and kindred and members of families who have during the rest of the year been scattered assemble to join at the table of the head of the family.[43]

Easter was similarly marked by community rituals rooted in the family. It was common throughout the north for instance to exchange 'Easter eggs' between children of friendly families.[44]

Across the face of England there was a host of festivals and rituals, often local and parochial, sometimes based in religious observances, sometimes in the rhythms of the agricultural or urban seasons, which provided a framework for collective and individual leisure. This is not the place to study ritual itself, but merely to draw attention to the degree to which many of those rituals were family-based. Indeed in large measure the family was *the* basic element, the fundamental constituency, from which the fabric of social recreation – and worship – was created. It was, of course, a reciprocal process; it was expected that many of these celebrations would, for instance, strengthen the bonds of family life. And when the need to reinforce the structure of family life became so commonly broadcast in early Victorian England in religious and social terms it was because those concerned felt that family life was being tested by the transforming pressures of social and economic change. However artificial the calls now seem – from Dickens, the churches or whatever – to shore up the family, they reflect not only on the assumed destructiveness of change at large but, more fundamentally, on the existence and importance of the family unit itself. The ideal of the Victorian middle-class family and domestic life ought not to distract us from the fundamental fact of the continuing durability and strength of the nuclear family in English society throughout these years. This ought not to be so surprising, however, for the English family had, after all, been the most resilient and well-tested institution throughout centuries of social transformation.

The family survived, intact, the continuing process of change shaped by urbanization and industrialization. For wide sectors of plebeian life, these years involved a bruising and difficult experience. But it is surely significant that what was most bitterly resented by working people were the harmful consequences of social change on their families. Whichever aspect of urban plebeian life we care to examine in these years it was the accentuation of family distress that provided the central and recurring complaint against urban life. Changes in the nature of illness and disease, in patterns and rhythms of employment, changes in the dwelling place, changes in the treatment of the poor, and the harmful effects on learning (however minimal), all these and more were felt, not as mere abstract trends, but as material deteriorations within the experiences of the plebeian family. And yet for all that – for all the ailments, the harsh working and living conditions, for all the ignorance and more – the plebeian family remained much the same in the mid nineteenth as in the mid eighteenth century. It was an apparently immutable object surrounded by the most fundamental of changes. Indeed, there is a plausible case to be made that what provided the strongest, unbroken links in those continuities in social behaviour in these years was the family unit. Those forms of human activity which continued almost unaltered throughout these years – whether the rituals, the pleasures or others – were those most closely related to, or a function of, family life. Although the majority of families were, by the mid nineteenth century, urban-based, their elemental qualities were timeless.

Conclusion: continuity and change

By 1851 the population was almost 18 millions. Fifty years earlier it was a mere 9 millions, a growth which, in the words of Kitson-Clark, created 'a community of a completely different order from what had existed in England and Wales before 1800'.[1] This claim notwithstanding, the widespread continuities in key areas of social behaviour and organization are very striking indeed. Much of the nation's foodstuffs and clothing, and the marketing of those goods, changed remarkably little in these years. Equally and perhaps more surprising, the working lives of the English town dweller remained traditional in a host of major occupations. It is true that advancing industrialization was actively transforming whole sectors of the labour force into work units of a size, nature and style quite unknown to earlier generations (excepting perhaps certain slave societies). None the less the large scale factory system so typical of industrial society belongs to a later period. True, some of the most glaring industrial abuses developed in these years and were subjected to detailed political scrutiny followed by early attempts to curb the cruder excesses. But a mere glance at this endemic feature of English experience as late, say, as 1900 provides a useful antidote to the view that the mid century English had begun to eliminate the worst of their social excesses.

The political changes initiated in the 1820s and 1830s by reforming legislation laid the basis for subsequent amelioration of urban government, rudimentary social policy and even political representation. But this was of limited relevance and benefit to those armies of labouring people who seemed, even at mid century, to have gained so little for their efforts and for the toils of their forbears. This is not to deny the existence of certain material gains by working people but rather to suggest that such gains were limited, partial – often transient – and generally difficult to appreciate when set against the improving fortunes of their social betters. Similarly, it would be hard to claim that, when confronted by the evidence of mid century poverty, this changing society held out the prospects of minimizing and limiting the worst forms of deprivation.

And yet for all this, contemporaries were widely agreed that mid century marked a decisive shift in social and political matters, an almost definable lurch of the English people into more peaceable and more tranquil moods and behaviour. The ungovernable people of a century before had been transmuted, or so it seemed, into a social entity which could be governed and controlled, nationally and locally, with an assurance and confidence unknown to earlier ages. Of course this tranquillity was in many respects deceptive and was to be fractured at a number of points later in the nineteenth and twentieth centuries. None the less it is hard to deny the peace that had, by the end of our period, settled on the nation. It is unlikely that this peaceableness was a direct result of a material gains for, whatever their form, they were, by general agreement, fairly small scale, recently won and often precariously held. Nor is it plausible that the increasingly effective mid century policing systems can in themselves provide a sufficient explanation, although in certain regards the police undoubtedly played a role in securing local control.

In many respects these years form a transitional phase in English urban culture, a period in which for all the manifest upheavals that so bemused contemporaries, many traditional customs and patterns of behaviour were able to accommodate themselves, rather than succumb, to overt political hostility, or to the accidental assaults of circumstance. This was particularly striking in the case of recreational culture. In the capital and in new industrial towns the lower orders tried to enjoy themselves in the manner of their forbears. Increasingly, however, this was within the ever-tighter confines of local regulation and propertied resistance. Old customs were undoubtedly under attack and in many respects on the wane. But many survived.

By 1851 there had been changes in abundance. Among the middle classes there had evolved a range of professional organizations and skills which were eventually to become a notable feature of modern urban life. But even here, by mid century, the foundations had only recently been laid and it was not until the second half of the century that the Victorian professions were to blossom. Moreover, even by 1851, the embryonic professions continued to be hemmed in by an older mentality which resisted and distrusted the advance of the professions. The middle-class professions rarely addressed themselves to working-class needs or interests except perhaps in the fields of worship and education. More and more professionally trained clergymen went out to minister to the manifold needs of urban working people. Similarly, a new breed of teachers took their skills to their deprived contemporaries through a host of voluntary organizations. But, again, it was to be much later in the century that the newly-developed teaching profession effectively confronted the great mass of the people, thanks to the gradual establishment of compulsory education.

Some major improvements in the urban habitat were unquestionably effected. Rudimentary municipal control over the traditional 'nuisances' of town life (of water, sanitation, crime, the sick and the dead) had begun to impose a clear sense of order on urban life. Once more, however, we have only to consider the volume of complaint against all of these problems (excepting perhaps burial) later in the century to appreciate the continuing difficulties of urban life. The process of urban and population change continued (and indeed accelerated) in the second half of the century with the inevitable accumulation and continuation of these age-old problems.

It would be wrong to describe the changes in these years as mainly physical or practical. There were marked transformations in a number of important mentalities. The early Victorians viewed themselves in quite a different light from their forbears. The perceptions of society divided, if not always neatly, into classes which were determined by economics, moulded by historical and political pressures and which articulated separate (if unequal) aspirations and outlooks, had become widespread by mid century. At almost every turn the English divided up their social and economic life by class; at work, at home, on the trains, in the cemeteries, and even in the ideology that was embedded in a great deal of contemporary literature. The English had come to view social class as normal and proper. Indeed there were innumerable English people proud to advocate the benefits and attractions of society divided into classes. The analogies with the division of labour (which seemed to be the basis for contemporary English economic pre-eminence) were obvious and repeated; society functioned best, most efficiently, with each person and family playing their accepted role within the social class to which they had been allotted by birth or circumstance. Naturally enough this is not to claim that the language of class was always on the lips of contemporaries but rather that the acceptance of and belief in social class had become a widespread phenomenon by mid century, though it was arguably to become even more pronounced and contentious later in the century.

At every turn, English urban life by the mid nineteenth century seems, for all its undoubted innovations, to have been in a transitional phase. Contemporaries and historians have, quite properly, been initially impressed by the magnitude of population and urban growth. But the human and physical expansion of early nineteenth-century England – that tide of advancing humanity and brickwork so beloved of contemporary cartoonists – often obscures as much as it reveals. The surprise of contemporaries at the seismic and physical upheavals around them has, naturally enough, bequeathed to us an abundance of evidence about the breaks, the discontinuities, with an older and apparently rapidly disappearing society. It is now possible to see a different and in some respects a contrary image; of social behaviour which proved extraordinarily resiliant and durable in the teeth of major urban change. Indeed it has been one of the most significant contributions of social historians over the past twenty years to reveal the remarkable and diverse continuities in English urban life. Even in 1851 what is perhaps most striking is the degree to which English people – of all social classes – conducted their social lives much like their forbears. The English people remained to a remarkable degree, and despite the unmistakable changes in the nation's physical and economic contours, resolutely wedded to a host of traditional and deeply rooted customs and forms of behaviour.

References

Introduction

1 W. G. East, 'England in the 18th century', in H. C. Darby (ed.), *Historical Geography of England before 1800* (Cambridge 1961 edn), p. 464.
2 Quoted in P. J. Corfield, *The Impact of English Towns, 1700–1800* (Oxford 1981), p. 2.
3 W. O. Henderson (ed.), *J. C. Fischer and his Diary of Industrial England, 1814–1851* (London 1966), p. 161.
4 Corfield, *The Impact of English Towns*, p. 168.

Chapter 1 People and towns

1 William Hutton, *The History of Birmingham* (London 1835 edn), p. 34.
2 M. D. George, *Hogarth to Cruickshank. Social Change in Graphic Satire* (London 1967), p. 180.
3 Neil Tranter, *Population Since the Industrial Revolution* (London 1973), pp. 41–2.
4 R. D. Lee and R. S. Schofield, 'British Population in the 18th century', in R. Floud and D. McCloskey (eds.), *The Economic History of Britain*, 2 vols. (Cambridge 1981), Vol. 1, p. 31.
5 Tranter, *Population*, p. 52; H. Carter, 'Towns and Urban Systems, 1730–1900', in R. A. Dogshon and R. A. Butlin (eds.), *A Historical Geography of England and Wales* (London 1978), p. 370.
6 Elizabeth Gaskell, *Mary Barton* (1848) (1970 edn), p. 39.
7 Carter, 'Towns and Urban Systems. . .', pp. 371–4.
8 H. L. Jones, 'Agriculture, 1700–1780', in Floud and McCloskey, *The Economic History of Britain*, p. 78.
9 C. W.Chalklin, *The Provincial Towns of Georgian England* (London 1974), pp. 32–5.
10 W. K. East, 'England in the 18th Century', in H. C. Darby (ed.), *Historical Geography of England before 1800* (Cambridge 1961 edn), p. 492.
11 Chalklin, *Provincal Towns*, pp. 39–40.

12 Quoted in East, 'England in the 18th century', p. 500.
13 ibid., and Chalklin, *Provincial Towns*, pp. 47–51.
14 John K. Walton, *The English Seaside Resort, A Social History, 1750–1914* (Leicester, 1983), ch. 3.
15 ibid., and John K. Walton, *The Blackpool Landlady* (Manchester 1978), ch. 2.
16 Keith Wrightson, *English Society, 1580–1680* (London 1982), p. 128.
17 Tranter, *Population*, pp. 46–7.
18 R. Lawton, 'Population and Society, 1730–1900' in Darby (ed.), *Historical Geography*, p. 318.
19 Pamela Horn, *The Rural World, 1780–1850* (London 1980), p. 14.
20 Dorothy Wordsworth's Diary, in J. Aiken, *English Diaries of the 18th Century* (London 1944), pp. 99, 100, 102.
21 Henry Mayhew, *London Labour and the London Poor,* 4 vols. (1861) (New York 1968 edn), vol. 1, p. 2.
22 Raphael Samuel, 'Comers and Goers', in H. J. Dyos and M. Wolff (eds.), *The Victorian City*, 2 vols. (London 1973), ch. 5.
23 Quoted in M. D. George, *London Life in the 18th century* (London 1965 edn), p. 117.
24 Francis Sheppard, *London, 1808–1870. The Infernal Wen* (London 1971), pp. 1–6.
25 E. A. Wrigley and R. S. Schofield, *The Population History of England, 1541–1871. A Reconstruction* (London 1981), pp. 227–8.
26 John Aiken, *A Description of the Country from 30 to 40 miles round Manchester* (London 1795), pp. 192–4.
27 Enid Gauldie, *Cruel Habitations. A History of Working Class Housing, 1780–1918* (London 1974), ch. 4.
28 ibid., p. 74.
29 ibid., p. 93.
30 Quoted in David Vincent, *Bread, Knowledge and Freedom* (London 1981), p. 55.
31 ibid.

Chapter 2 Suffering people: illness and death in the urban world

1 Neil Tranter, *Population Since the Industrial Revolution* (London 1973), p. 53.
2 Neil Tranter, 'The Labour Supply, 1780–1860', in R. Floud and D. McCloskey (eds.), *The Economic History of Britain*, 2 vols. (Cambridge 1981), vol. 1, p. 213.
3 E. A. Wrigley and R. S. Schofield, *The Population History of England 1541–1871. A Reconstruction* (London 1981), pp. 667–70.
4 E. A. Wrigley, *Population and History* (London 1969), p. 165.
5 Tranter, 'Labour Supply', pp. 214–15.
6 John Burnett, *Plenty and Want. A Social History of Diet in England from 1815 to the Present Day* (London 1979 edn), p. 73.

7 F. B. Smith, *The People's Health, 1830–1910* (London 1979), pp. 65–9, 105–6.
8 ibid., pp. 66, 58.
9 ibid., 81–5, 90–1, 99, 104.
10 ibid., pp. 106–7, 136–7.
11 ibid., pp. 144–5, 239, 244–5.
12 ibid., p. 288; F. F. Cartwright, *A Social History of Medicine* (London 1977), p. 123.
13 John K. Walton, *The English Seaside Resort, A Social History, 1750–1914* (Leicester 1983), pp. 143–4.
14 Smith, *The People's Health*, pp. 330–1.
15 M. J. Durey, *The First Spasmodic Cholera Epidemic in York, 1832* (York, Borthwick Papers, no. 46, 1974), p 2.
16 ibid., p. 28.
17 Quoted in ibid., pp. 28–9.
18 Quoted in Smith, *The People's Health*, p. 230.
19 John Snow, *On the Communication of Cholera* (London 1854, New York edn 1936), pp. 135–6.
20 Durey, *The First Spasmodic Cholera*, p. 14.
21 Smith, *The People's Health*, p. 198.
22 C. H. Hume, 'The Public Health Movement', in J. T. Ward (ed.), *Popular Movements, c. 1830–1850*, (London 1970), pp. 184, 186–7.
23 Quoted in ibid., p. 191.
24 Harold Perkin, *The Origins of Modern English Society, 1780–1880* (London 1972 edn), pp. 169–71.
25 Smith, *The People's Health*, pp. 249–50.
26 Perkin, *Origins*, p. 121.
27 Tranter, 'Labour Supply', pp. 214–15.
28 N. Parry and J. Parry, *The Rise of the Medical Profession* (London 1973), p. 104.
29 M. J. Durey, *The Return of the Plague* (Dublin 1979), pp. 180, 183.
30 ibid., pp. 208–9, 213.
31 Smith, *The People's Health*, pp. 346, 349–50.
32 ibid., pp. 109–11.
33 Virginia Berridge, *A bad use of opium; working class drug use in 19th century England* (paper given to Social History Conference, Chester December 1981); and Virgina Berridge, *Opium and the People* (London 1981).
34 ibid., Cartwright, *A Social History of Medicine*, p. 111.
35 Smith, *The People's Health*, pp. 343–5.
36 ibid., pp. 333–42.
37 *Parliamentary Papers, 1852–1899. Alphabetical Index* (London) vol. 1, p. 167.
38 *Accounts and Papers* (1863), vol. XLVL, 'Cemeteries'.
39 G. A Walker, *On the Past and Present State of the Intramural Burying Places* (London 1851), pp. 9, 6.

40 G. A. Walker, *Gatherings from the Grave Yards* (London 1839), p. iii.
41 Alfred E. Hargrave, *The Baneful Custom of Interment in Towns and the Present State of the York Grave Yards* (York 1847), pp. 5–7.
42 ibid., pp. 11–18.
43 *Yorkshire Gazette* (22 November 1851), p. 5; (4 May 1853), p. 7.
44 *The Victoria County History, The History of Yorkshire*, P. M. Tillott (ed.), *The City of York* (London 1961), p. 466.
45 James Steven Curl, *A Celebration of Death* (London 1980), pp. 211–12.
46 Quoted in ibid., p. 212.
47 ibid., pp. 233, 245.
48 P. H. J. H. Gosden, *The Friendly Societies in England, 1815–1875* (New York edn 1967), p. 7.
49 Sir John Simon (1852), in E. Royston Pike, *Human Documents of the Victorian Golden Age* (London 1967), p. 286.
50 R. J. Morris, *Cholera, 1832* (London 1976), pp. 105–6; James Walvin, 'Dust to Dust. The Celebration of Death in Victorian England', *Historical Reflections* (spring, 1983).
51 E. Royston Pike, *Human Documents of the Industrial Revolution* (London 1966), p. 354.
52 Sidney and Beatrice Webb, *English Poor Law History,* Part II, *The Last 100 Years,* vol. 1 (London 1929), p. 446; vol 2, p. 1054.
53 Alison Yapp, *London Burial 1831–1850* (undergraduate thesis, Ealing College of Higher Education 1980), pp. 8, 24–8, 30–1.
54 Burial Fees', *Accounts and Papers*, XLV, (1854–5), pp. 55;22;48.30.
55 Yapp, *London Burial.*
56 Quoted in David Vincent, *Bread, Knoweledge and Freedom* (London 1981), p. 56.

Chapter 3 Working Lives

1 J. T. Ward, *The Factory System*, 2 vols. (Newton Abbott 1970), vol. 2, p. 17.
2 E. J. Hobsbawm, *Industry and Empire* (London 1968), ch. 3.
3 Michael Anderson, *Family Structure in Nineteenth Century Lancashire* (Cambridge 1971), pp. 39–40.
4 ibid., p. 105.
5 ibid., p. 117.
6 Robin M. Reeve, *The Industrial Revolution, 1750–1850* (London 1971), p. 64; J. F. C. Harrison, *Quest for the New Moral World* (New York 1969), p. 154.
7 D. A. Farnie, *The English Cotton Industry and the World Market, 1815–1876* (Oxford 1979), pp. 13, 18–20, 24–5.
8 ibid., p. 35.
9 Hobsbawm, *Industry*, p. 47
10 P. J. Corfield, *The Impact of English Town 1700–1800* (Oxford 1981), pp. 96–7.

11 James B. Jeffreys, *Retail Trading in Britain, 1850–1950* (Cambridge 1954), p. 5.
12 Duncan Bythell, *The Sweated Trades* (London 1975), p. 145.
13 ibid., pp. 124–5, 132, 146.
14 Gareth Stedman Jones, *Outcast London* (London 1971 edn), p. 26.
15 Bythell, *Sweated Trades*, pp. 147, 151.
16 John Burnett, *Plenty and Want. A Social History of Diet in England from 1815 to the Present Day* (London 1979 edn), pp. 54–5.
17 ibid., pp. 187–8, 20–1.
18 S. G. Checkland, *The Rise of Industrial Society in England 1815–1885* (London 1971 edn), pp. 215–16; Reeve, *The Industrial Revolution*, p. 65.
19 John Burnett, *Useful Toil* (London 1974), p. 135.
20 ibid., p. 136.
21 Pamela Horn, The *Rural World, 1780–1850* (London 1980), p. 251.
22 Quoted in Pamela Horn, *The Rise and Fall of the Victorian Servant* (Dublin 1975), p. 18.
23 F. F. Cartwright, *A Social History of Medicine* (London 1977), pp. 50–7.
24 Harold Perkin, *The Origins of Modern English Society, 1780–1880* (London 172 edn), pp. 254–5.
25 ibid., p. 122; A. D. Gilbert, *Religion and Society in Industrial England* (London 1976), p. 130.
26 Gilbert, *Religion and Society*, p. 27.
27 ibid., p. 134.
28 G. A. Campbell, *The Civil Service in Britain* (London 1965), p. 16.
29 E. W. Cohen, *The Growth of the British Civil Service, 1780–1939* (London 1965 edn), chs. 1–5.
30 J. Lawson and H. Silver, *A Social History of Education in England* (London 1973), pp. 257–8, 299–300.
31 ibid., pp. 300–2.
32 See for instance, 'The Army' in L. H. Gann and Peter Daignan, *The Rulers of British Africa*, (London 1978).
33 Perkin, *Origins*, pp. 58–60.
34 ibid., pp. 135–8.
35 P. K. O'Brien and S. Engerman, 'Changes in Income and its distribution during the industrial revolution', in R. Floud and D. McCloskey (eds.), *The Economic History of Britain*, 2 vols. (Cambridge 1981), vol. 1, p. 166.
36 ibid., pp. 169–71.
37 Perkin, *Origins*, p. 145.
38 Bythell, *Sweated Trades*, p. 150.
39 Burnett, *Useful Toil*, pp. 156, 160–3.

Chapter 4 The language of class

1 Roy Porter, *English Society in the 18th century* (London 1982), p. 68.

2 E. P. Thompson, 'Eighteenth Century English Society; class struggle without class?', *Social History,* 3 no. 2 (May 1978), p. 163.

3 Porter, *English Society*, p. 81. On the broader question of land ownership, see David Cannadine, *Lords and Landlords: The Aristocracy and the Towns, 1774–1967* (Leicester 1980).

4 Quoted in James Walvin, *Beside the Seaside. A Social History of the Popular Seaside Holiday* (London 1978), p. 16.

5 Lillian M. Penson, 'The West India Interest in the 18th century', in R. Mitchison (ed.), *Essays in Eighteenth Century History* (London 1966), p. 1; see also Douglas Hall, *A Brief History of the West India Committee* (Barbados 1971).

6 Derek Fraser, *Urban Politics in Victorian England* (Leicester 1976), pp. 25–6.

7 ibid., p. 91.

8 Pamela Horn, *The Rural World, 1780–1850* (London 1980), p. 152.

9 Harold Perkin, *The Origins of Modern English Society, 1780–1880* (London 1972 edn), pp. 40, 271–2.

10 Edward Royle and James Walvin, *English Radicals and Reformers, 1760–1848* (Brighton 1982), pp. 60–2.

11 Asa Briggs, 'The Language of "class" in early nineteenth century England', in Asa Briggs and John D. Saville (eds.), *Essays in Labour History* (London 1967).

12 Quoted in Clive Emsley, *British Society and the French Wars 1793–1815* (London 1979), pp. 50–1.

13 Quoted in ibid., p. 78.

14 Quoted in ibid., p. 80

15 ibid., p. 176.

16 Peter Burke, *Sociology and History* (London 1980), pp. 60–7.

17 John Brewer, 'English Radicalism in the Age of George III', in J. G. A. Pocock (ed.), *Three British Revolutions; 1641, 1688, 1776* (Princeton, NJ 1980), pp. 323–67.

18 Briggs, 'Language', p. 59.

19 Asa Briggs, *The Age of Improvement,* (London 1962 edn), p. 313.

20 Briggs, 'Language', p. 61.

21 *The Charter* (29 December 1839), in Edward Royle, *Chartism* (London 1980), pp. 106–7.

22 Briggs, *The Age of Improvement*, p. 320.

23 Royle, *Chartism* pp. 76–7.

24 Royle and Walvin, *English Radicals,* p. 158.

25 ibid., p. 169.

26 Quoted in Briggs, 'Language', p. 71.

27 David Cannadine, 'Residential Differentiation in Nineteenth Century Towns', in J. H. Johnson and C. G. Pooley (eds.), *The Structure of Nineteenth Century Cities* (London 1982), pp. 238, 245.

28 Asa Briggs, *Victorian Cities* (London 1968 edn), pp. 33–4.

Chapter 5 The poor

1 Clive Emsley, *British Society and the French Wars 1793–1815*, ch. 3.
2 David Vincent, *Bread, Knowledge and Freedom* (London 1981), p. 52.
3 J. H. Treble, *Urban Poverty in Britain* (London 1979), pp. 86–7.
4 Derek Fraser, *The Evolution of the British Welfare State* (London 1973), pp. 32–3; M. E. Rose, *The Relief of Poverty* (London 1981 edn), pp. 17–18.
5 F. O'Gorman, *The Emergence of the British Two Party System, 1760—1832* (London 1982), p. 87.
6 Quoted in Fraser, *Evolution*, p. 38.
7 Rose, *Relief of Poverty*, p. 17.
8 E. J. Hobsbawm and George Rudé, *Captain Swing* (London 1973 edn).
9 Rose, *Relief of Poverty*, p. 17.
10 Fraser, *Evolution*, pp. 40–1, 48.
11 Rose, *Relief of Poverty*, p. 17.
12 Pamela Horn, *The Rural World, 1780–1850* (London 1980), ch. 4
13 Rose, *Relief of Poverty*, p. 17.
14 ibid., p. 20.
15 E. P. Thompson and Eileen Yeo (eds.), *The Unknown Mayhew* (London 1973 edn), p. 3.
16 ibid., pp. 22–4, 32.
17 Quoted in Fraser, *Evolution*, p. 123.
18 John K. Walton, *The English Seaside Resort, A Social History, 1750–1914* (Leicester 1983), ch. 5.

Chapter 6 Learning and believing

1 J. Lawson and H. Silver, *A Social History of Education in England* (London 1973), pp. 192–3.
2 R. W. Malcolmson, *Life and Labour in England 1700–1780* (London 1981), p. 95.
3 Lawson and Silver, *A Social History*, p. 278; Pamela Horn, *The Rural World, 1780–1850*, (London 1980), p. 138.
4 Anne Digby and Peter Searby, *Children, School and Schooling in 19th century England* (London 1981), pp. 3–5.
5 Lawson and Silver, *A Social History*, p. 278.
6 John Brewer, *Party Ideology and Popular Politics at the Accession of George III* (Cambridge 1976), pp. 140, 142–3, 146–7.
7 Quoted in Edward Royle and James Walvin, *English Radicals and Reformers, 1760–1848* (Brighton 1982), pp. 61, 59–60.
8 R. Brimley Johnson (ed.), *The Letters of Hannah More* (London 1925), p. 174.
9 Thomas W. Laqueur, 'Working Class Demand and the Growth of English Elementary Education, 1750–1850', in L. Stone (ed.), *Schooling and Society* (London 1978), ch. 8.

10 ibid., p. 192.
11 ibid., p. 197
12 David Vincent, *Bread, Knowledge and Freedom* (London 1981), p. 95.
13 ibid., p. 97.
14 ibid., p. 99.
15 ibid., pp. 129–30.
16 Thomas W. Laqueur, *Religion and Respectability. Sunday Schools and Working Class Culture, 1780–1850* (London 1975), pp. 244–5.
17 Keith Wrightson, *English Society, 1580–1680* (London 1982), p. 185.
18 A. D. Gilbert, *Religion and Society in Industrial England* (London 1976), p. 110.
19 ibid., pp. 107–8.
20 ibid., pp. 112, 125.
21 W. R. Ward, *Religion and Society in England, 1790–1850* (London 1972), p. 252; G. Kitson Clark, *The Making of Victorian England* (London 1966 edn), pp. 48–9.
22 J. F. C. Harrison, *The Early Victorians, 1832–1851* (London 1971), pp. 123–4.
23 Edward Royle, *Victorian Infidels* (Manchester 1973).
24 Quoted in K. S. Inglis, *Churches and the Working Classes in Victorian England* (London 1964), p. 20.
25 ibid., p. 121; Neil Tranter, 'The Labour Supply, 1780–1860', in R. Floud and D. McCloskey (eds.), *The Economic History of Britain*, 2 vols. (Cambridge 1981), vol. 1, pp. 210–11.
26 James Obelkevich, *Religion and Rural Society; South Lindsey, 1825–1875* (Oxford 1976) p. 319.
27 Hugh McLeod, *Class and Religion in the Late Victorian City* (London 1974), p. 279.
28 Edward Royle, 'Religion in York, 1831–1981', in C. B. Feinstein (ed.), *York, 1831–1981* (York 1981), pp. 208–10, 226.
29 McCleod, *Class and Religion*, p. 281.
30 Quoted in Kitson Clark, *Making of Victorian England,* p. 164.
31 McCleod, *Class and Religion*, p. 283.

Chapter 7 Challenging the systems: radicalism and the development of mass politics

1 David Geggus, *Slavery, War and Revolution* (Oxford 1981).
2 This is an accusation levelled at this series. See John Bossy, 'Social England', *Encounter* (September–October 1982).
3 Edward Royle and James Walvin, *English Radicals and Reformers, 1760–1848* (Brighton 1982).
4 David Vincent, *Bread, Knowledge and Freedom* (London 1981), p. 114.
5 Samuel Bamford, *Passages in the Life of a Radical* (London 1839), p. 11.
6 ibid., pp. 11–12.
7 Royle and Walvin, *English Radicals*, pp. 128–31.

8 On Church and King mobs, see John Stevenson, *Popular Disturbances in England, 1700–1870* (London 1979), pp. 137–42.
9 See 'Introduction' in James Walvin (ed.), *Slavery and British Society, 1776–1848* (London 1982).
10 ibid.; and Seymour Drescher, 'Public Opinion and the Destruction of British Colonial Slavery', in Walvin (ed.), *Slavery and British Society*, pp. 22–48.
11 M. J. Durey, *The Return of the Plague*, (Dublin 1979); R. J. Morris, *Cholera, 1832* (London 1976).
12 Royle and Walvin, *English Radicals*, p. 148.
13 On riots see Stevenson, *Popular Disturbances*.
14 Royle and Walvin, *English Radicals*, p. 180.

Chapter 8 The people subdued: law and order in an urban setting

1 The expression is from John Brewer and John Styles (eds.), *An Ungovernable People* (London 1980).
2 John Brewer, *Party Ideology and Popular Politics at the Accession of George III* (Cambridge 1978).
3 John Stevenson, *Popular Disturbances in England 1700–1870* (London 1979), ch. 2.
4 ibid., p. 83.
5 *The Manchester Directory*, 2 vols. (1788), vol. 1, pp. 51–2, 56 (in Manchester Central Library).
6 Quoted in Stevenson, *Popular Disturbances*, p. 90.
7 ibid., p. 163.
8 Quoted in ibid., p. 174.
9 Details taken from *Manchester Chronicle* (January–March 1793).
10 For Pitt's personal fears about associations see his letters to Henry Dundas, November 1792, in *Pitt Papers*, William L. Clement Library, University of Michigan, Ann Arbor.
11 Quoted in Clive Emsley, *British Society and the French Wars 1793–1815* (London 1979), p. 174.
12 Stevenson, *Popular Disturbances*, pp. 218–22.
13 M. J. Durey, *The First Spasmodic Cholera Epidemic in York, 1832* (York, Borthwick Papers, no. 46, 1974), and *The Return of the Plague* (Dublin 1979).
14 E. J. Hobsbawm and George Rudé, *Captain Swing* (London 1973 edn); Pamela Horn, *The Rural World, 1780–1850* (London 1980), pp. 88–95.
15 Gwyn A. Williams, *The Merthyr Rising* (London 1978), p. 18.
16 Quoted in Edward Royle, *Chartism* (London 1980), pp. 91–2.
17 ibid., pp. 25–6, 97.
18 *The Manchester Directory*, vol. 1, pp. 51–2, 56.
19 David Jones, *Crime, Protest, Community and Police in the Nineteenth Century* (London 1982).
20 Brian Harrison, *Drink and the Victorians* (London 1971).
21 Jones, *Crime, Protest*, p. 21.

22 *Manchester Mercury* (11 March 1788).
23 James Walvin, *A Child's World* (London 1982), pp. 186–92.
24 ibid., ch. 11.
25 Chris Cook and John Stevenson, *The Longman Handbook of Modern British History, 1714–1980* (London 1983), p. 126.
26 M. Ignatieff, *A Just Measure of Pain: the Penitentiary in the Industrial Revolution* (New York 1978).
27 Royle, *Chartism*, p. 94.
28 ibid., p. 97.
29 Geoffrey Best, *Mid-Victorian Britain, 1851–75* (London 1971), p. 229.

Chapter 9 People's pleasures

1 Peter Burke, *Popular Culture in Early Modern Europe* (London 1978); E. P. Thompson, 'Rough Music: Le charivari anglais', *Annales,* 27 (1972).
2 Keith Wrightson, *English Society,* 1580–1680 (London 1982), pp. 167–70.
3 Roy Porter, *English Society in the 18th Century* (London 1982), ch. 6; P. J. Corfield, *The Impact of English Towns, 1700–1800* (Oxford 1981), ch. 4.
4 The expression is used by Francis Place, quoted by M. D. George, *London Life in the 18th Century* (London 1965 edn), p. 18.
5 David Cooper, *The Lesson of the Scaffold* (London 1974), pp. 7–9, 20–1.
6 James Turner, *Reckoning with the Beast* (Baltimore 1980).
7 Hugh Cunningham, *Leisure in the Industrial Revolution* (London 1980), pp. 22–4.
8 Pamela Horn, *The Rural World, 1780–1850* (London 1980), pp. 172–3.
9 Quoted in R. W. Malcolmson, *Popular Recreation in English Society, 1700–1850* (Cambridge 1973), p. 153.
10 Horn, *Rural World,* p. 181.
11 James Walvin, *The People's Game* (London 1975).
12 Quoted in Cunningham, *Leisure,* p. 81.
13 Quoted in ibid., p. 84.
14 Peter Bailey, *Leisure and Class in Victorian England* (London 1978), pp. 9–10.
15 R. D. Storch, 'The Plague of Blue Locusts', *International Review of Social History,* 20 (1975); David Philips, ' "A New Engine of Power and Authority". The Institutionalisation of Law Enforcement in England 1780–1830', in V. A. C. Gatrell, Bruce Lenman and Geoffrey Parker (eds.), *Crime and the Law* (London 1980), p. 189.
16 Frederick Engels, *The Condition of the Working Class in England in 1844* (London 1968 edn), p. 102.
17 Cunningham, *Leisure,* p. 94.
18 David Russell, *The Popular Musical Societies of the Yorkshire Textile District, 1850*–1914 (DPhil. thesis University of York 1979).
19 ibid., pp. 33–5.
20 ibid., pp. 40,37.

21 Cunningham, *Leisure*, p. 104.
22 Brian Harrison, *Peaceable Kingdom. Stability and Change in Modern Britain* (Oxford 1982), ch. 4.
23 Porter, *English Society*, pp. 245–6.
24 John K. Walton, *The English Seaside Resort, A Social History, 1750–1914* (Leicester 1983).
25 James Walvin, *Beside the Seaside. A Social History of the Popular Seaside Holiday* (London 1978), p. 39; J. B. Harley, 'England c. 1850', in H. C. Darby (ed.), *A New Historical Geography of England after 1600* (Cambridge 1961 edn), pp. 289–90.
26 Porter, *English Society*, p. 247; A. B. Granville, *Spas of England*, 2 vols. (London 1841).
27 Quoted in Walvin, *Beside the Seaside*, pp. 38–9.
28 James Walvin, 'Children's Pleasures', in John K. Walton and James Walvin (eds.), *Leisure in Britain, 1780–1939* (Manchester 1983).
29 I. Stickland, *The Voices of Children, 1700–1914* (Oxford 1973), p. 154.
30 E. P. Thompson, and Eileen Yeo (eds.), *The Unknown Mayhew* (London 1973 edn), pp. 339–40.

Chapter 10 Rituals and ceremonies

1 James Obelkevich, *Religion and Rural Society, South Lindsey, 1825–1875* (Oxford 1976), pp. 82–6.
2 Derek Fraser, *Urban Politics in Victorian England* (Leicester 1976), p. 116.
3 Asa Briggs, *Victorian Cities* (London 1968 edn), p. 138.
4 David Cannadine, 'Civic Ritual and the Colchester Oyster Feast', *Past and Present*, no. 94 (February 1982).
5 John Brewer, 'The Wilkesites and the law, 1763–74', in John Brewer and John Styles (eds.), *An Ungovernable People* (London 1980), pp. 134–5.
6 Leon Radzinowicz, *A History of Criminal Law and its Administration from 1750*, 4 vols. (London 1948), vol. 4, p. 344.
7 Albert Goodwin, *The Friends of Liberty* (London 1979), ch. 10.
8 Mark Judd, 'The oddest combination of town and country; Popular culture and the London fairs, 1800–1860', in John K. Walton and James Walvin (eds.), *Leisure in Britain, 1780–1939* (Manchester 1983), ch. 1.
9 ibid., pp. 16, 20.
10 ibid., p. 25. See also E. P. Thompson, 'Rough Music: Le charivari anglais', *Annales*, 27 (1972); and Peter Burke, *Popular Culture in Early Modern Europe* (London 1978).
11 Robert Poole, 'Oldham Wakes', in John K. Walton and James Walvin, (eds.), *Leisure in Britain 1780–1939* (Manchester 1983), p. 83.
12 ibid., p. 72.
13 ibid., p. 87.
14 David Russell, 'Popular musical culture and popular politics in the York-

shire textile districts, 1880–1919', in Walton and Walvin (eds.), *Leisure*, p. 109.

15 James Walvin, 'Dust to Dust. The Celebration of Death in Victorian England, *Historical Reflections* (spring 1983).

16 ibid.

17 'List of cemeteries', *Accounts and Papers*; XLVI (1863).

18 John Brand, *Observations on the Popular Antiquities of Great Britain*, 3 vols. (London 1848–9; New York edn 1970).

19 ibid., vol. 1, p. vii.

20 William Hone, *The Year Book* . . . (London 1838).

Chapter 11 Family Life

1 Roy Porter, *English Society in the 18th Century* (London 1982), p. 159.

2 Karl Marx and Frederick Engels, *The Communist Manifesto*, (1848, 1967 edn), p. 100.

3 Peter Lazlett, *Family Life and illicit love in earlier generations* (Cambridge 1978), pp. 12–13.

4 Porter, *English Society*, p. 163.

5 Jeffrey Weekes, *Sex, Politics and Society* (London 1981).

6 Porter, *English Society*, pp. 281–2.

7 F. B. Smith, *The People's Health 1830–1910* (London 1979), ch. 1.

8 Edmund Burke, 'Sketch of a Negro Code' (1792), in *The Works of the Right Honourable Edmund Burke*, 16 vols. (London 1826), vol. 9.

9 V. G. Kiernan, *The Lords of Human Kind* (London 1972 edn).

10 E. Wrigley and R. S. Schofield, *The Population History of England, 1541–1871. A Reconstruction* (London 1981), see index of 'nuptiality'.

11 Beatrice Webb, *My Apprenticeship* (London 1971 edn), p. 133.

12 *How to be a Lady* (London 1850).

13 ibid., pp. 165–6, 364.

14 Mrs Valentine (ed.), *The Girl's Home Companion* (London 1895), p. 752.

15 *Children in the Potteries*, 2 vols. (Staffordshire County Council 1975), vol. 1 p. 9.

16 Anne Digby and Peter Searby, *Children, School and Schooling in 19th Century England* (London 1981), p. 46.

17 John Pearson to Jane Munby, 9 February 1799, *Munby Papers*, York City Archives, Acc. 54, 194–201.

18 Catherine Hall, 'Gender Division and class formation in the Birmingham Middle Class, 1780–1850', in R. Samuel (ed.), *People, History and Socialist Theory* (London 1981), pp. 166–7.

19 *The Girl's Companion* (London 1868), p. 723.

20 On these and other rituals at a later date see Robert Roberts, *The Classic Slum* (London 1971 edn).

21 Marjorie Cruikshank, *Children and Industry* (Manchester 1981), p. 31.

22 David Vincent, *Bread, Knowledge and Freedom* (London 1981), p. 109.

23 ibid., p. 110.
24 ibid., p. 131.
25 Michael Anderson, *Family Structure in Nineteenth Century Lancashire* (Cambridge 1971), pp. 112–13.
26 James Walvin, *A Child's World* (London 1982), ch. 4.
27 Anderson, *Family Structure,* pp. 122–3.
28 ibid., pp. 124, 134–5.
29 ibid., p. 88.
30 ibid., p. 76.
31 Quoted in Walvin, *Child's World,* p. 64.
32 Vincent, *Bread,* p. 60.
33 ibid., pp. 41–2.
34 ibid., pp. 46–7.
35 Anderson, *Family Structure,* pp. 132–5.
36 'The Death of Little Children' in A. Symons (ed.), *Essays of Leigh Hunt* (London 1903), p. 44.
37 Vincent, *Bread,* p. 56.
38 ibid., p. 59.
39 John Brand, *Observations on the Popular Antiquities of Great Britain,* 3 vols. (London 1848–9, New York edn 1970), vol. 2, pp. 87–196.
40 ibid., p. 59.
41 James Walvin, 'Dust to Dust. The Celebration of Death in Victorian England', *Historical Reflections* (spring 1983).
42 J. A. R. Pimlott, *The Englishman's Christmas* (Brighton 1978 edn); R. W. Malcolmson, *Popular Recreations in English Society, 1700–1850* (Cambridge 1973), pp. 26–7.
43 Quoted in Pimlott, *Englishman's Christmas*, p. 94.
44 Brand, *Observations*, vol. 1, p. 168.

Conclusion: continuity and change

1 G. Kitson Clark, *The Making of Victorian England* (London 1966 edn), p. 66.

Index

www.ingramcontent.com/pod-product-compliance
Ingram Content Group UK Ltd.
Pitfield, Milton Keynes, MK11 3LW, UK
UKHW011007180926
475701UK00004B/613